Mohamed Abdel-Raheem
Abeer Saleh Alabdallah

The important insects that infest wheat crops in Syria and Egypt

Mohamed Abdel-Raheem
Abeer Saleh Alabdallah

The important insects that infest wheat crops in Syria and Egypt

Noor Publishing

Imprint

Cover image: www.ingimage.com

Publisher:
Noor Publishing
is a trademark of
Dodo Books Indian Ocean Ltd. and OmniScriptum S.R.L publishing group

120 High Road, East Finchley, London, N2 9ED, United Kingdom
Str. Armeneasca 28/1, office 1, Chisinau MD-2012, Republic of Moldova, Europe
Managing Directors: Ieva Konstantinova, Victoria Ursu
info@omniscriptum.com

Printed at: see last page
ISBN: 978-620-9-31665-4

The important insects that infest wheat crops and methods of controlling them in both Syria and Egypt

By

Prof. Dr. Mohamed Abdel-Raheem

Dr. Abeer Saleh Alabdallah

(2026)

The important insects that infest wheat crops and methods of Controlling them in both Syria and Egypt

Prof. Dr. Mohamed Abdel-Raheem Ali Abdel-Raheem

Professor of Entomology (Biological Control)

Pests & Plant Protection Department

Agricultural and Biological Research Institute

National Research Centre.

Dr. Abeer Saleh Alabdallah

PhD in Agricultural Engineering

Department of Plant Protection

Affiliated with the Directorate of Institutes at Al-Furat University, Syria

In the name of ALLAH

The Most Gracious, the Most Merciful, All praise is due to God, first and last, praise befitting His majesty and the greatness of His gifts, for He facilitated and aided us in completing this scholarly work, and enabled us to bring it to light after effort, research, and perseverance. To Him be all thanks for His blessings, and to Him belongs all favor for what He inspired and taught.

My colleague, Prof. Dr. Mohamed Abdel-Raheem, and I are pleased to present to researchers, interested parties, and students this book entitled "The Most Important Economic Insects Affecting Wheat Crops and Methods of Controlling Them in Both Syria and Egypt." To be a scientific reference that contributes to identifying the most important insect pests that threaten this strategic crop, and presents their biological characteristics, damages and ways to combat them in a documented scientific manner that serves students, agricultural engineers and researchers in the field of plant protection.

The importance of this book stems from the urgent need for a scientific resource that combines academic accuracy with practical application, and helps to spread awareness of wheat pests and ways to reduce their losses, in support of food security, and in service of the agricultural sector in our Arab region, especially in the environments of both Syria and Egypt, given their agricultural weight and long history in wheat cultivation.

As we conclude this endeavor, we ask God Almighty to make this work beneficial knowledge, a continuous act of charity, and purely for His sake, and that it be in the service of God and in the service of knowledge, seeking thereby to advance knowledge, benefit people, and contribute to the advancement of agricultural scientific research.

And ALLAH is the source of all success.

Content list

Subject	Page
Introduction	7
The importance of wheat in Syria	7
The importance of wheat in Egypt	8
Benefits of wheat plant	8
First: it is rich in essential nutrients:	8
Second: supporting cardiovascular health	9
Third: preventing diabetes and improving blood sugar control	9
Fourth: supporting digestive health	9
Fifth: Antioxidants and Active Plant Compounds	9
Sixth: The Role of Wheat in the Environment and Sustainable Agriculture	10
Insect pests affecting wheat crops in both Syria and Egypt	11
First: The most important insects that infest wheat in Syria	11
Second: The most important insects that infest wheat in Egypt	12
Sunn pest	13
Aphids	20
Fall armyworm	24
Cereal ground beetle	30
Cutworms	35
Cereal leaf miner	41
Wheat Stem Sawfly	49
Hessian fly	54

Cereal Leaf Beetle	58
Winter grain chafer	62
Wheat and Grass Thrips	66
The concept of the economic threshold in wheat	70
First: *Eurygaster integriceps*	70
Secondly: wheat aphids (*Rhopalosiphum padi-, Sitobion avenae*)	71
Third: The rural armyworm, *Spodoptera frugiperda*	71
Fourth: Cereal seedling moth, *Zabrus tenebrioides*	71
Fifth: Cutworms, Agrotis spp.	72
Sixth: Cereal Leafminer, *Syringopais temperatella*	72
Seventh: Wheat sawfly, *Cephus pygmaeus*	72
Eighth: Hess fly, *Mayetiola destructor*	72
Ninth: Cereal leaf beetle, *Oulema melanopus*	73
Tenth: Winter grain, Anisoplia spp.	73
Eleventh: Wheat Thrips (Thrips spp.)	73
Table (1): containing the scientific and common names of insect pests that infest wheat	74
Table (2): containing the scientific and common names of the most important crops mentioned in the book.	75
References	76

The important insects that infest wheat crops and methods of controlling them in both Syria and Egypt

Introduction

The grass family (Poaceae) is one of the most important plant families in global agriculture, as it includes staple grains that feed the world's population and support food security, such as wheat (*Triticum* spp.), corn (*Zea mays*), and barley (*Hordeum vulgare*), which represent the largest part of global food demand.

Global cereal production, including wheat, is estimated to reach record levels in the 2025/2026 season, according to the Food and Agriculture Organization (FAO, 2025a) forecast. In the Arab world, wheat is a key crop in food security policies, as countries seek to boost local production and reduce dependence on imports through agricultural support programs and expanding cultivated areas, especially in the face of population pressures and climate and water challenges (FAO, 2025b).

The importance of wheat in Syria

Wheat is a strategic crop in Syria, as it was previously able to meet a large part of local food needs, but production has declined sharply in recent years due to drought, weak rainfall, and high temperatures that have affected cultivated areas and productivity (FAO GIEWS, 2025). According to the FAO GIEWS reports, Syria's wheat import needs for the 2025/2026 season are estimated at about 3 million tons, which is about 70% higher than the five-year average (FAO GIEWS, 2025).

The area planted with wheat in the 2024/2025 season also decreased to about 1.1 million hectares, with an estimated production of only about 922,000 tons compared to previous seasons with higher production, indicating a clear challenge in achieving national food security (Shafaq, 2025).

The importance of wheat in Egypt

In the Arab Republic of Egypt, wheat is the primary food crop and plays a pivotal role in national food security, particularly in the production of subsidized bread, which is the staple food for the majority of the population (USDA FAO, 2025a).

Agriculture data show that the area planted with wheat and the varieties used contributed to raising domestic production for 2025 to levels close to the historical average, supported by government policies to provide subsidized seeds and encourage local farming (USDA FAO, 2025b). Nevertheless, Egypt still partially relies on imports to cover the gap between consumption and domestic production, as a result of high population demand compared to productive growth (FAO GIEWS, 2025b).

Benefits of wheat plant

First: it is rich in essential nutrients:

Wheat and its products, such as whole wheat, contain high amounts of complex carbohydrates, proteins, fiber, vitamins (such as B1, B3, and B6), and minerals (such as iron, magnesium, and zinc) that are essential for the body. These nutrients support daily energy and promote vital bodily functions (Shewry, 2009; FAO, 2011).

Second: supporting cardiovascular health:

The dietary fiber in whole wheat reduces levels of LDL (bad) cholesterol and improves blood vessel health, thus lowering the risk of heart disease (Anderson et al., 2000; Jenkins et al., 2002).

Third: preventing diabetes and improving blood sugar control:

The fiber in whole wheat helps slow glucose absorption, reducing post-meal blood sugar spikes and improving insulin sensitivity (Ludwig, 2002; et al., 2018; Chiavaroli).

Fourth: supporting digestive health:

The dietary fiber in whole wheat improves bowel movements and reduces constipation, and it also supports the growth of beneficial bacteria in the colon, thus contributing to overall digestive health (Slavin, 2003; Jones, 2014).

Fifth: Antioxidants and Active Plant Compounds:

The outer layer of wheat contains phenolic and flavonoid compounds that act as antioxidants, reducing inflammation and protecting cells from damage caused by free radicals (Xu & Chang, 2007; Wang et al., 2009).

Sixth: The Role of Wheat in the Environment and Sustainable Agriculture:

Wheat cultivation contributes to improved land use and is an important crop for crop rotation, reducing the spread of diseases and pests and enhancing soil fertility (FAO, 2017; Shewry & Hey, 2015).

Despite the great importance of cereals and wheat in food security, the actual productivity of the crop is affected by various biological factors, most notably insect pests. It is well known that insects such as the sunn pest and other insect pests can cause significant production losses and may greatly reduce grain yield and quality if not managed through integrated pest management (IPM) programs (Wikipedia, 2025).

Scientific studies show that such insects can affect not only plant growth, but also post-harvest grain characteristics, including nutritional and quality losses, making understanding the insect life cycle crucial.

The factors contributing to their spread and the manifestations of infestation in plants form an important scientific basis for designing effective management strategies, enabling the reduction of production losses and the improvement of post-harvest grain quality (Wikipedia, 2025b). The following chapters of this book will be dedicated to reviewing the most important of these insects that infest wheat in both Syria and Egypt, along with a comprehensive analysis of the scientific methods available for their control.

Insect pests affecting wheat crops in both Syria and Egypt

Wheat is one of the crops most exposed to insect pests in the agricultural environments of the Middle East, and the spread of these pests leads to a significant decrease in productivity and a decline in grain quality, especially in Syria and Egypt, which are among the most important wheat-growing regions in the area.

The severity of infestation and the type of pests vary according to climate and farming systems, but a number of pests are common, widespread, and economically impactful at the field and storage levels.

First: The most important insects that infest wheat in Syria

The sunn pest *Eurygaster integriceps* is the most dangerous pest to wheat in Syria, as it causes a significant decrease in productivity as a result of feeding on the leaves and spikelets, in addition to its direct impact on flour quality due to the salivary enzymes that cause "protein breaks" in the grains (FAO, 2022; Syrian Ministry of Agriculture, 2023).

Extensive infestations of aphids such as *Sitobion avenae* and *Rhopalosiphum padi* are also recorded, which weaken growth and are a major vector for wheat viruses such as yellow barley dwarf virus (CIMMYT, 2021; SARC, 2022).

In recent years, the fall armyworm *Spodoptera frugiperda* has been observed in some areas of northern Syria, causing damage to the vegetation and a decrease in plant productivity (FAO 2023; IPPC, 2022).

The hess fly, *Mayetiola destructor*, which weakens the internodes and causes empty spikes (USDA, 2021; ICARDA, 2020), is also widespread, in addition to the wheat stem borer, *Cephus pygmaeus*, which leads to stem breakage and reduced yield (Davis *et al.*, 2020; Arab Journal of Plant Protection, 2022).

No less important are storage pests such as grain weevils, flour beetles and grain moths, which contribute to significant post-harvest losses when stored poorly (FAO, 2020; APS, 2021).

Second: The most important insects that infest wheat in Egypt

In Egypt, the fall armyworm *S. frugiperda* tops the list of pests affecting wheat since its introduction to the country, causing leaf erosion and a decline in vegetative growth during the early growth stages (Egyptian Ministry of Agriculture, 2023; FAO, 2022).

Aphid infestations of *Sitobion avenae* and *Schizaphis graminum* are also widespread in the Delta and Upper Egypt regions, and their danger stems from their role in transmitting dwarfing viruses that significantly limit grain filling (ARC, 2021; CIMMYT, 2022).

The wheat fly, *Mayetiola destructor*, is an important pest that causes seedling weakness and the appearance of underdeveloped ears (USDA, 2021; Plant Protection Research Institute, 2022). Infestations of the wheat sawfly, *Cephus pygmeus*, are also recorded in the northern Delta, causing premature stem breakage and associated yield loss (Arab Journal of Plant Protection, 2023; ICARDA, 2020). Wheat leafhoppers, which contribute to the transmission of viruses in dry and semi-dry environments (FAO, 2021; ARC, 2023), also appear, in addition to storage pests that cause qualitative and quantitative losses in stored grains (FAO Grain Storage, 2020). We will discuss all of the above in detail.

Sunn pest

Classification

Scientific name: *Eurygaster integriceps* Put

Family: Pentatomidae (Stink-smelling bugs)

Order: Hemiptera (Schuh & Slater, 1995)

Origin

Southern Europe and the Mediterranean basin are the native habitat of this insect, where it is adapted to temperate climatic conditions (Brown, 1962; EPPO, 2013).

Distribution

The sunn pest is widespread throughout most of the Middle East and West Asia and is considered one of the most serious economic pests of cereal crops, particularly wheat and barley, in countries such as Syria, Iraq, Turkey, Iran, and Egypt. It is a significant economic pest of wheat and barley farms. (CABI, 2022; FAO, 2018)

Hosts

The sunn pest feeds mainly on wheat (*Triticum aestivum*) and barley (*Hordeum vulgare*), in addition to some wild grasses which are a food source for nymphs in the early stages before they move to cultivated crops (Paulian & Popov, 1980).

Morphological Description

The sunn pest has an oval body, tending towards brown or gray in color, and has a foul odor that is released when crushed. The length of the adult insect usually ranges between 12–15 mm, and its forewings are characterized by partially covering the body, while the hindwings are transparent and used for flight during migration periods (Brown, 1962; EPPO, 2013) (Figure 1).

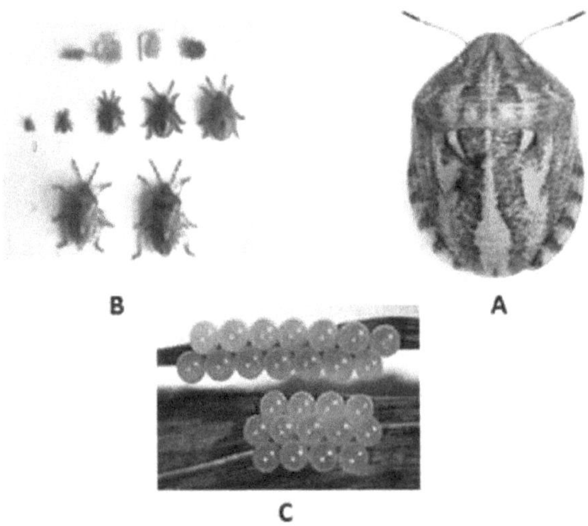

Figure (1): A. The adult insect – B. The insect's life cycle – C. Egg laying

Damage and Symptoms:

The sunn pest damages wheat and barley ears by sucking the sap from the grains during their milky stage, leading to a reduction in the grain's specific weight, resulting in shrinkage, lighter weight, lower starch content, and reduced protein (gluten) content in the flour.

This renders the resulting flour unsuitable for dough and bread making, and sometimes leads to complete crop loss and delayed ripening. The infection leaves a distinctive mark on the affected grains, appearing as small sunken spots on their surface (Hariri et al., 2000; FAO, 2018) (Figure 2).

A B

Figure (2): A. Symptoms of infection on the ears; B. The difference between infected and healthy ears.

Life Cycle

It spends the winter as an adult insect in the surrounding hills and highlands, then migrates to the wheat fields in late March and early April to mate. The female lays eggs from late March until the end of April in groups of 8-15 eggs on the underside of plant leaves.

The eggs hatch after 7-10 days into nymphs that feed first on wild plants and then move to wheat and barley. After reaching maturity, they transform into adults that feed on the ears of grain at midday, then transform into fully grown insects that feed voraciously on wheat ears.

It is in the milk stage, after which it stops feeding and migrates to vineyards and mountains to secure protection from direct sunlight during the summer. The sunn pest has one generation per year (Brown, 1962; Paulian & Popov, 1980; EPPO, 2013) (Figure 3), and is characterized by a dual migration behavior:

1. Spring migration: Adult insects move from the wintering areas to the fields during mid-March, and continue for about 20 days to feed and reproduce.
2. Summer migration: It begins in mid-June, when the new generation of insects abandons the fields towards higher areas to enter a long dormancy of up to 8–9 months under fallen leaves or soil (Kriticos, 2017; FAO, 2018).

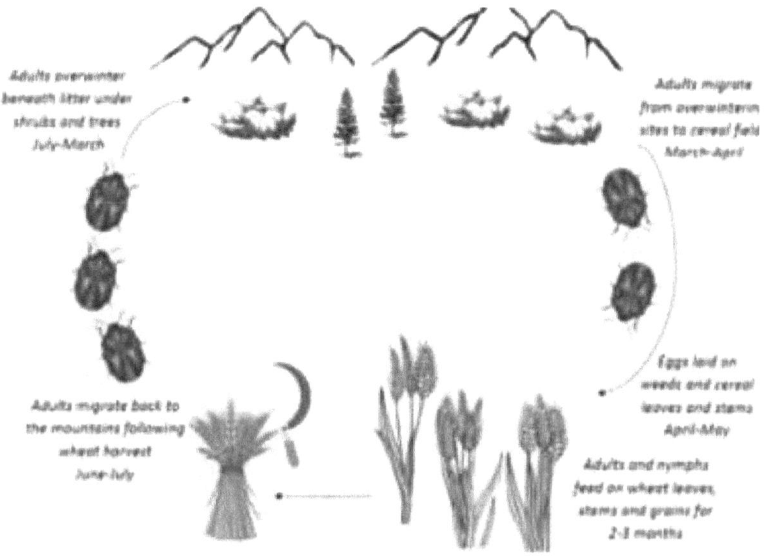

Figure (3): shows the life cycle stages of the sunn pest.

Control Methods

First: Cultural Control:

Planting early-maturing wheat varieties, thus allowing for early harvesting, reduces the period of grain exposure to attack. Proper tillage and improved plant nutrition through fertilization also increase plant strength and resistance to the insect.

Also, planting early-maturing varieties of wheat and barley reduces sunn pest damage (FAO, 2018; EPPO, 2013).

Secondly: chemical control:

Using chemical pesticides in cases of severe, high-density infestations can reduce insect populations and limit their spread, provided that the recommended timing and dosages within integrated pest management programs are followed.

Phosphorus or carbamate pesticides can be applied directly to the host plant, such as:

Trichlorofon at a rate of 2 kg/dunum – Phenithione – Carbaryl (Cefin). Dipterex 80% at a rate of 30 g/dunum (Brown, 1962; FAO, 2018).

Third: Biological Control:

Several natural parasites are available that can be used within integrated pest management programs to limit the spread of the insect without affecting the agricultural environment. In Syria, the eggs of the sunn pest are attacked by several parasites, the most important of which are:

The insect *Telenmus vassilieui* can kill 50% of the sunn pest's eggs and color them black. The fly *Phasia subcoleoptrata* L. attacks the sunn pest when it opens its wings and prepares to fly, quickly laying its eggs under the wings. The hatched larva feeds on the insect's internal organs and kills it.

There are also several bird species that attack the adult insect and nymphs. Additionally, there are several wasps that parasitize the eggs. Furthermore, there are several flies belonging to the family Tachnidae (Paulian & Popov, 1980; CABI, 2022). (Figure 4)

Figure (4): A. *Telenmus vassilieui* - B. *Phasia subcoleoptrata*

Aphids

Scientific name: *Rhopalosiphum padi* Linnaeus

Order: Hemiptera

Family: Aphididae

Common name: Oat aphid or papyrus aphid, one of the most important wheat pests and a major source of yellow sickness virus (BYDV) (Ahdb, 2023).

Scientific name: *Sitobion avenae* Fabricius

Order: Hemiptera

Family: Aphididae

Common name: English grain aphid, which attacks wheat, barley and oats, and reduces yield and ear quality (Agrobiotop, 2023).

Distribution and Spread both species are found globally in temperate regions where cereals are grown (Ahdb, 2023). R. padi and S. avenae constitute the majority of aphid populations in wheat fields (JPPP, 2023).

Morphological Description of *R. padi*:

The adult insect is 1–2 mm long, olive-green or pale in color, with two dark-colored appendages on either side of the fifth abdominal segment called corniches (Ahdb, 2023). The insect exhibits winged and wingless forms depending on environmental conditions (Ahdb, 2023). It reproduces asexually in summer and sexually on trees in autumn (Ahdb, 2023) (Figure 5).

Figure (5): an aphid.

Damage to wheat caused by the species *R. padi*:

The nymph and adult insect suck sap from the plant's leaves, causing stunted growth and poor development (UC IPM, 2023). Its greatest harm is that it acts as a vector for BYDV (Barley Yellow Dwarf Virus) and also for wheat, significantly reducing yields (Ahdb, 2023).

The life cycle of *R. padi*:

Winged forms migrate from trees to wheat fields in the spring (Ahdb, 2023). Rapid asexual generations develop on the grain during the season (Ahdb, 2023). Some return to the trees in the autumn to lay overwintering eggs (Ahdb, 2023).

Morphological description of S. avenae: The adult insect is 2–3 mm long, green or yellow in color, sometimes with brown or purple spots (Agrobiotop, 2023). It exists in winged and wingless forms depending on the environment, reproducing asexually in spring and summer and sexually in autumn (Wikipedia, 2023).

Damage to wheat caused by *S. avenae*:

Both the nymph and adult insect suck sap from the leaves and ears, leading to a decrease in the number and weight of grains (Encyclop'Aphid, 2023). It also secretes honeydew, which promotes the growth of sooty mold fungi and thus inhibits photosynthesis (Encyclop'Aphid, 2023). Severe infestations can result in losses of up to 2500 kg/hectare (Encyclop'Aphid, 2023).

The life cycle of S. avenae: The insects appear in the spring on wheat and complete successive asexual generations during the growth of the ears (Agrobiotop, 2023). They can move between fields and weeds depending on environmental conditions (Agrobiotop, 2023). (Figure 6)

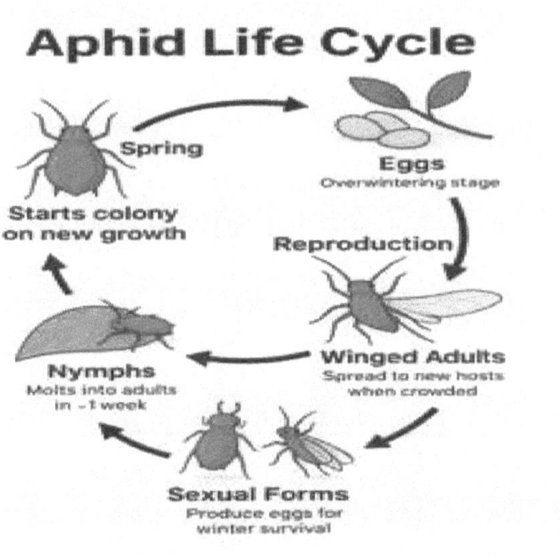

Figure (6): the stages of the aphid's life cycle.

Control Methods

First: Field Monitoring:

Calculate the average aphid population per plant to determine intervention (Purdue, 2023).

Second: Natural Enemies:

Several natural enemies prey on aphids, such as ladybugs, parasitic wasps, and tarantulas (JPPP, 2023).

Plowing and weed control can reduce migrating aphid populations (UC IPM, 2023).

Pesticides:

Used when the economic threshold is exceeded, and the appropriate pesticide must be chosen to avoid the development of resistance in insects (PubMed, 2020).

Fall armyworm

Scientific name: (J.E. Smith, 1797) *Spodoptera frugiperda*

Family: Noctuidae

Order: Lepidoptera

It is known globally as the fall armyworm because of its ability to move in large groups when heavily infested. (FAO, 2022; CABI, 2023)

Origin

This pest originates from the warm regions of the Americas, where it was endemic for centuries before spreading beyond the continent. Numerous reports indicate that tropical and subtropical climates provided a suitable environment for its high-density reproduction. (Goergen *et al.*, 2016; FAO, 2022)

Distribution

Within a few years, the fall armyworm transformed from a pest confined to America to a global pest, with its presence recorded in many places such as Africa, the Middle East (including Syria, Iraq, and Egypt), South and East Asia, and Southern Europe. Its rapid spread is attributed to its great ability to fly and migrate long distances. (CABI, 2023; Eppo, 2021)

Host Plants

This pest is known for its wide range of host plants, as it can attack more than 300 plants, including: maize (the most susceptible host), wheat, barley, millet and sorghum, rice, cotton, and a number of vegetables. It may infect wheat, especially at the seedling and early vegetative stages (FAO, 2022; CABI, 2023).

Morphological Description

The adult insect: A brown butterfly with forewings decorated with pale spots. It has a great ability to fly, and this feature is one of the most important reasons for its widespread distribution. (CABI, 2023).

The larva:

The larva is the damaging stage of this insect. Its color tends towards green in the early stages and then turns dark brown. On the head is an inverted Y-shaped mark, which is a distinctive feature for differentiating between similar species. At the end of the body are four dark spots arranged in a square shape. The fully grown larva may reach a length of about 40 mm. (Capinera, 2017; EPPO, 2021) (Figure 7)

The Pupa: It forms in the soil inside a small cavity, and its color is reddish-brown.

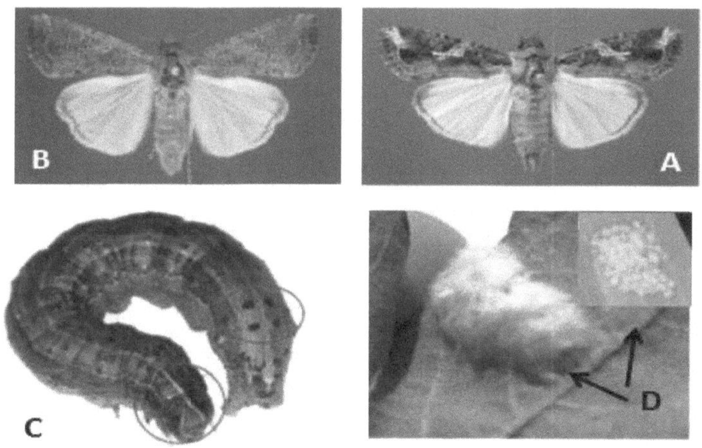

Figure (7): the insect's life cycle stages: A. Male - B. Female - C. Larva - D. Egg-laying stage

Damage and Symptoms of Infestation on Wheat

Infestation begins with larval attacks on seedlings, usually by gnawing at leaf tips, leaving irregular holes. Early infestation can lead to stunted growth. It also causes damage to the leaf tissue, as the larvae leave dark-colored droppings between the leaves, and torn lines appear on the upper surface. If the larval population increases, the following occurs:

Slow plant growth, reduced ear formation, a marked decrease in grain production, and in severe cases, damage can exceed 20–30%. (FAO, 2022; EPPO, 2021) (Figure 8)

Figure (8): the symptoms of armyworm infestation.

Life Cycle

The armyworm is highly prolific, completing 8–10 generations annually in warm environments. Eggs are laid in clusters covered with bristles by the female, who can lay around 2,000 eggs per season. The eggs hatch into larvae, which undergo six instars, the longest and most damaging stage.

After completing its growth, it returns to the soil to become a pupa, spending its metamorphosis period in the soil. Then the adult butterfly emerges; it has a relatively short lifespan but is highly mobile, and is the main source of transmission of the infection over distances. (CABI, 2023; Capinera, 2017)

Control methods Integrated Pest Management (IPM)

The management of the fall armyworm relies on the application of the Integrated Pest Management (IPM) concept, which aims to reduce insect populations below the economic threshold using a combination of agricultural, biological, mechanical and chemical means, while minimizing environmental impact and health risks (FAO, 2018; EPPO, 2022).

First: Agricultural Control:

Sound agricultural practices are among the most effective means of limiting the spread of the fall armyworm, as they directly affect its ability to survive and reproduce. Furthermore, regulating and standardizing planting dates within a single agricultural area contributes to reducing the availability of plants of different ages. This limits the continuity of food sources for the larvae and reduces the severity of infestation (Prasanna et al., 2018).

Crop rotation with crops that are not hosts for the insect also breaks its life cycle and reduces the accumulation of its numbers between growing seasons (EPPO, 2022). Deep plowing after harvest helps to expose the pupae present in the soil to weather conditions and natural enemies.

This leads to a decrease in its survival rate. Furthermore, controlling grassy weeds surrounding the field reduces the alternative hosts the insect relies on during periods when the main crop is absent (Day *et al.*, 2017).

Second: Mechanical and physical control:

Mechanical methods play a supporting role, especially in the early stages of infestation. Manual collection of eggs and larvae is an effective procedure in small spaces and nurseries, as it helps to reduce the insect population density before it reaches the more damaging advanced larval stages (FAO, 2018).

Pheromone traps are also used to monitor the activity of adult butterflies and determine the timing of generation emergence, which helps in selecting the most appropriate time to apply various control measures (Prasanna *et al.*, 2018).

Third: Biological control:

Biological control is considered one of the most sustainable methods due to its role in maintaining ecological balance. This control involves utilizing natural natural enemies, such as general predators (spiders and ground beetles) and specialized parasites, particularly egg parasites of the genus *Trichogramma.*

Larval parasites such as *Cotesia marginiventris* contribute to reduced hatching and larval growth rates (Kenis *et al.*, 2019).

Microbial agents are also used successfully, most notably *Bacillus thuringiensis*, which affects the digestive system of young larvae, as well as insect-specific polynuclear viruses.

These fungi cause high mortality rates under suitable conditions. Insect-pathogenic fungi, such as *Beauveria bassiana* and *Metarhizium anisopliae*, are an effective option, especially in high-humidity environments (Harrison *et al.*, 2019).

Fourth: Chemical Control:

Chemical pesticides are used within integrated pest management programs only when necessary, and after the infestation level has exceeded the economic threshold. Some newer pesticides, such as anthranilamides and spinosins, show high efficacy against early larval stages.

When using these chemicals, it is essential to adhere to the recommended dosages and alternate between different chemical groups to minimize the risk of resistance development, while observing the required safety intervals to protect human health and the environment (EPPO, 2022; Prasanna *et al.,* 2018).

Fifth: Integrated Pest Management (IPM):

Integrated pest management is the most effective framework for managing the fall armyworm, as it is based on a balanced combination of agricultural, biological, mechanical and chemical means, with reliance on periodic monitoring and appropriate decision-making in a timely manner.

This approach has proven remarkably successful in reducing the economic damage caused by the insect in many agricultural areas around the world (FAO, 2018; Kenis *et al*., 2019).

Cereal ground beetle

Scientific name: *Zabrus tenebrioides* (Goeze, 1777)

Order: Coleoptera

Family: Carabidae

Also known as: Cereal seedling chewer or winter grain beetle. (CABI, 2022; EPPO, 2021)

Origin

Reports indicate that the pest's original habitat is southeastern Europe and temperate regions around the Mediterranean basin, from where it later spread to western Asia and North Africa with the expansion of wheat and barley cultivation. (Backlund, 1986; CABI, 2022)

Distribution

The insect is found in: Turkey, Syria, Iraq, Eastern and Western Europe, the Caucasus, Iran, and parts of North Africa. It is noted to be more widespread in rainfed wheat areas and relatively cool, humid climates. (EPPO, 2021; CABI, 2022) Plant hosts (Host Range): Wheat is the primary host for the seedling chewer, and it also feeds on barley, oats and some grassy weeds in cool, moist soil (Backlund, 1986; CABI, 2022).

Morphological Description

Larva: (damaging stage) It possesses strong mandibles capable of gnawing seedlings. It is of the flattened type, with long thoracic legs. Its color is creamy white except for the head and thoracic segments, which are dark brown. It has two appendages at the end of the abdomen and can reach 18–22 mm in length when fully grown. Its body is relatively shiny and clearly segmented. (EPPO, 2021) Adult insect (beetle)

The adult insect: A shiny black beetle, relatively slow-moving and nocturnal, 12–16 mm long, with a narrow first thorax and a pitted posterior margin. The femur is black, while the tarsus and leg are red (CABI, 2022).

The pupa: Develops in an earth chamber within the soil (Figure 9).

Figure (9): the life stages of the grain seedling chewer: A. Adult insect - B. Larva

Damage and Symptoms

The adult insect feeds on the wheat ears and grains as soon as they form at the beginning of summer. The larvae, however, attack the leaves of the seedlings, pulling them into tunnels in the soil and chewing them, leaving only the fibers. They also feed on the roots.

A single larva consumes approximately 25–40 wheat plants until it reaches maturity, resulting in a significant reduction in yield and substantial economic losses. (FAO, 2020 ؛Backlund, 1986).

Characteristics of infestation:

Larvae are found in the soil at a depth of 2–5 cm, and seedlings disappear completely in small patches within the field. Some studies indicate that losses in cold years can reach up to 50% of the rainfed crop. (Backlund, 1986; FAO, 2020)

Life Cycle

Zabrus tenebrioides is a winter pest, with eggs laid in the soil during late summer. The larvae develop during autumn and enter partial overwintering, then resume activity at the beginning of spring. The larvae feed more frequently, causing greater damage. The pupae form in late spring within the soil.

The adult beetle emerges in the summer and then hides until the following season. The insect has one generation per year (univoltine). (CABI, 2022; EPPO, 2021) (Figure 10)

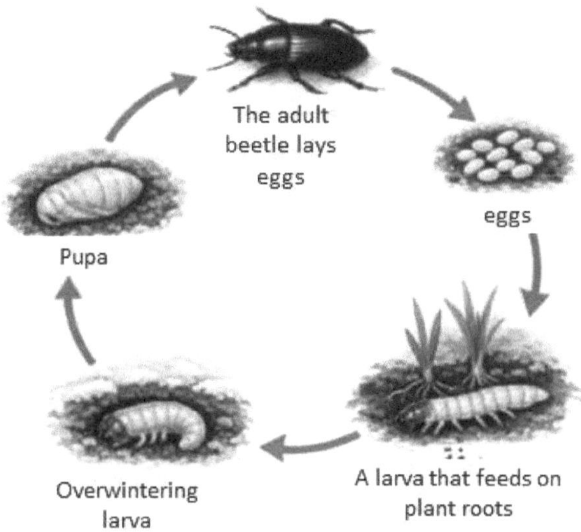

Figure (10): Life cycle of *Zabrus tenebrioides*

Integrated Pest Management (IPM) Control Methods

First: Agricultural Control:

Deep plowing after harvest to expose larvae and pupae to birds and sunlight. Reducing grass density around fields. Using slightly higher seeding rates in areas expected to be infested. (FAO, 2020)

Second: Biological control:

Natural predators from the Carabidae family feed on the eggs and larvae. Some parasitic species from the Braconidae family may attack the larvae, but they are not commercially effective. (Backlund, 1986)

Third: Chemical control: Treatment is recommended when 2–3 larvae/m^2 are observed.

With synthetic pyrethrin pesticides, chlorantranilproyl, ambda-cyhalothrin, and preferably applied at the beginning of the emergence of larvae in the fall. (EPPO, 2021; CABI, 2022)

Cutworms

Scientific name: *Agrotis ipsilon - Agrotis segetum*

Order: Lepidoptera

Family: Noctuidae

They are known as cutworm larvae, and most belong to the genus Agrotis. Some of the most important species that attack wheat include:

Agrotis ipsilon - Agrotis segetum (FAO, 2023 ؛CABI, 2022ز)

Origin

The original habitat of most Agrotis species likely extends across the Mediterranean, Asia, and North Africa, before spreading globally with the expansion of agriculture and cereals. (EPPO, 2021; CABI, 2022)

Distribution

The cutworm is a widespread global pest, found in: the Middle East (Syria, Iraq, Jordan), North Africa, Europe, Central Asia, and the Americas. It is classified as one of the most important ground pests of cereal crops worldwide. (FAO, 2023; USDA, 2022)

Host Range

The cutworm feeds on a wide variety of plants, most notably wheat, barley, maize, cotton, legumes, grasses, and dense weeds (CABI, 2022; El-Lakwah et al., 2019)

Morphological Description

The adult insect (butterfly): Grayish-brown with dark spots on the forewings, it is nocturnal and moves towards light.

The caterpillar: Thick and cylindrical, dark gray to blackish-brown in color, it can reach 40–45 mm in length when fully grown. When disturbed, it curls into a ring or a C-shape, a distinctive behavior for identification. (USDA, 2022; Capinera, 2018)

The pupa: It is reddish-brown in color and is found inside a cavity in the soil at a depth of 3–6 cm. (Capinera, 2018; EPPO, 2021) (Figure 11)

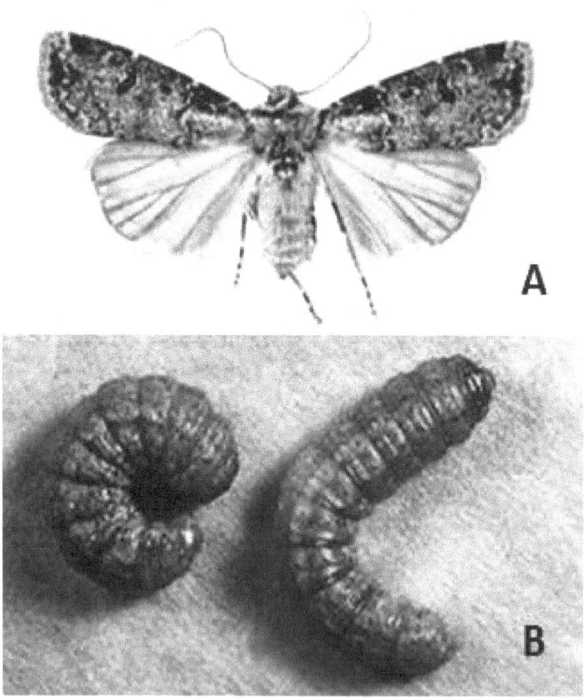

Figure (11): the life stages of the cutworm: A. Adult insect - B. Larva

Damage and Symptoms:

The larva attacks wheat by gnawing at seedlings from the soil surface, resulting in noticeable gaps in the planting rows. It also gnaws at young stems at night, causing the plant to wilt. In severe cases, losses can exceed 30–40% (FAO, 2023; El-Lakwah *et al.*, 2019).

One of the distinctive symptoms of this insect is the presence of C-shaped coiled larvae when digging up soil, cut and fallen plant remains. (Figure 12).

Figure (12): the symptoms caused by cutworm infection.

Life Cycle in Brief

Cutworms are among the most important insect pests affecting wheat crops in the early stages of growth. They are characterized by a complete metamorphic life cycle, including the egg, larva, pupa, and adult stages. The larval stages are the most dangerous due to the direct damage they cause to seedlings (Hill, 2012; Capinera, 2017).

It spends its winter in the fully grown larval stage and in some species in the pupal stage inside the soil at a depth that usually ranges between 5-15 cm, and sometimes under plant remains or grasses. In the spring, the adult insect (butterfly) emerges.
It is a nocturnal butterfly that feeds on flower nectar and is not considered an economically damaging stage, although it is responsible for egg-laying and the start of infestation (Hill, 2012; Pedigo & Rice, 2014). Then it mates, and the females begin laying eggs on: the soil surface, plant debris, the bases of wheat plants, or grassy weeds.

The eggs are spherical or nearly spherical, pearly white at first, then darken before hatching. The female lays hundreds of eggs during her lifetime, and the egg stage lasts several days depending on the temperature (Capinera, 2017; CABI, 2022). The eggs hatch into larvae (cutworms), which are the main damaging stage. The larvae often live hidden in the soil during the day.

It is active at night, gnawing at the stems of wheat seedlings at or slightly below the soil surface, causing the plant to be cut and fall (Hill, 2012; EPPO, 2021). The larva goes through several instars (usually 5–6), and the damage increases with the size of the larva. Infestations are more severe in fields with dense growth or those containing weeds (Capinera, 2017; Pedigo & Rice, 2014).

Upon reaching maturity, the larva transforms into a pupa within an earth chamber in the soil. The pupal stage is immobile and does not feed, and this stage lasts from several days to several weeks depending on environmental conditions. This stage is a transitional phase for tissue reorganization before the emergence of the adult insect (Hill, 2012; CABI, 2022).

Cutworms can have one or several generations per year depending on the species and climatic conditions. In temperate and warm regions, generations may overlap (EPPO, 2021; Capinera, 2017).

Cereal leaf miner

Scientific name: *Syringopais temperatella* L.

Order: Lepidoptera

Family: Scythrididae

Also known as: Oriental Cutter Wheat Butterfly or Oriental Soilworm. (CABI, 2022; EPPO, 2021)

Origin

It is believed to have originated in the Eastern Mediterranean region, particularly Syria, Turkey, Palestine, and Jordan, and its spread has since extended to parts of Iraq and Iran. (FAO, 2020; Kfir, 1983)

Geographic Distribution

The pest is mainly found in Syria, particularly in rainfed areas, northern and central Iraq, Turkey and southern Anatolia, Palestine, and Jordan, and represents one of the most serious pests of rainfed wheat in the region. (CABI, 2022; EPPO, 2021)

Host Range

Wheat is the primary host of this insect, but it also attacks barley and some grasses such as Avena spp. and Bromus spp. It prefers crops in arid and semi-arid regions. (Kfir, 1983; FAO, 2020)

Morphological Description

Adult: A small moth measuring 12–15 mm in length, with grayish-brown forewings and dark spots. It flies short distances and is active in the morning and evening. (EPPO, 2021; FAO, 2020)

Larva: Small, measuring 8–12 mm in length, with a creamy-gray color and a dark brown head.

Thin tunnels of silk and soil around the bases of plants. (Kfir, 1983; CABI, 2022)

Pupa: Transforms inside a silken cocoon among soil and plant remains. (Figure 13)

Figure 13 shows the life stages of the plantworm: A. Larva - B. Butterfly

Damage and Symptoms

This pest is more dangerous than common cropworms (Agrotis spp.) on rainfed wheat because:

First: It feeds inside tunnels at the base of the stem, and the larvae feed on the lower leaves and stem bases, and silken tunnels covered with soil are seen around the plant.

Second: It causes gradual yellowing of the plants, weak seedlings, stunted growth, and a significant decrease in the number of ears.

Third: It reduces productivity, with losses potentially reaching 40–60% in arid regions during epidemic years, particularly for rainfed wheat. (FAO, 2020; Kfir, 1983) (Figure 14)

Figure 14 shows the symptoms caused by infestation with the **Cereal leaf miner**

.

Life Cycle

The female lays her eggs singly or in small groups on plant debris or on the soil surface near wheat and barley seedlings, especially in poorly tilled fields. The eggs are small, oval-shaped, and yellowish-white when laid, gradually darkening before hatching.

The incubation period of the eggs usually extends from 7 to 14 days depending on the prevailing temperature and humidity (Al-Zyoud & Katbeh-Bader, 2011; EPPO, 2022). The eggs then hatch into larvae. The larvae are the economically damaging stage, as the larvae, immediately after hatching, head to the leaves of wheat or barley and burrow inside the leaf blade, forming distinctive longitudinal tunnels.
It feeds on the mesophyll, resulting in translucent spots or white lines that later turn brown and cause the leaves to dry out. The larva goes through several larval instars (usually 4–5), and the larval stage lasts from 30 to 45 days. In severe infestations, the larvae may cause stunted vegetative growth and reduced photosynthetic capacity of the plant (Kalash & Said, 2004; CABI, 2023).

This insect exhibits a double dormancy (summer and winter), where fully grown larvae enter a state of rest inside silken cocoons in the soil or among crop residues. This stage is an important adaptation for resisting adverse environmental conditions such as drought and high temperatures in summer or cold temperatures in winter.
The dormant stage may last for several months until suitable conditions are present for pupation (EPPO, 2022; FAO, 2018). Upon completion of dormancy, the larva

pupates inside a thin silken cocoon in the soil. The pupa is brown, and this stage typically lasts from 10 to 20 days, depending on climatic conditions.
This stage does not cause any direct harm to the plant (Al-Zyoud et al., 2015; Hill, 2008).

The adult butterflies emerge in the spring. They are small, have narrow wings, and are grayish-brown in color, which helps them camouflage themselves. The adults are mostly active in the evening and mate shortly after emerging. The females then begin laying eggs to start a new life cycle.

The insect has one generation per year in most of its distribution areas in the Middle East and Eastern Mediterranean (CABI, 2023; Kalash & Said, 2004). (Figure 15)

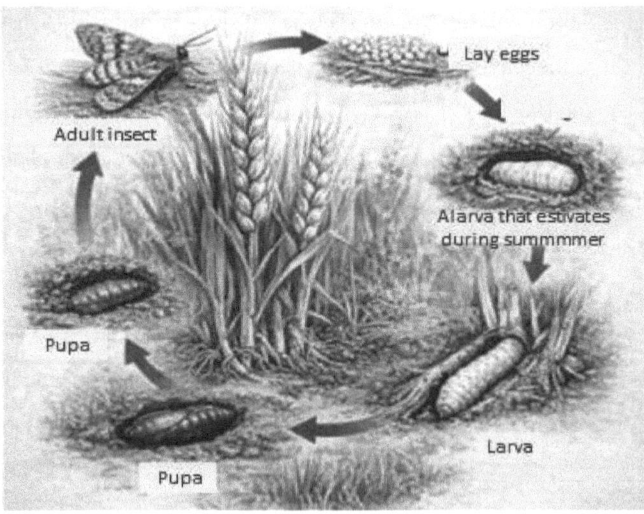

Figure 15 Life cycle *Syringopais temperatell*

Control methods Integrated Pest Management (IPM)

Control of the cereal leaf miner relies on Integrated Pest Management (IPM), since the larva lives most of its life inside the leaves or soil, making chemical control alone insufficient (EPPO, 2022; CABI, 2023).

First, agricultural control:

This is one of the most important methods of reducing infestation, as the insect spends a large part of its life cycle in the soil or crop residues. Deep plowing after harvest helps expose dormant larvae and pupae to unfavorable environmental conditions and natural enemies (Kalash & Said, 2004; FAO, 2018).

Removing the remains of the previous crop reduces egg-laying sites and larval survival during dormancy (EPPO, 2022). Crop rotation with non-grass crops (legumes such as beans and lentils) also limits the pest's persistence in the field (Hill, 2008; FAO, 2018), and planting at appropriate times reduces the synchronization of active larval stages with the sensitive stages of the wheat plant (CABI, 2023).

Second. Mechanical and physical control:

Such as leveling the land well reduces the cracks in which the larvae hide, and shallow hoeing in the early stages of wheat growth may kill some of the larvae located near the soil surface (Kalash & Said, 2004).

Third, biological control:

Biological enemies play an important role in naturally reducing pest populations, especially in low-pesticide farming systems. Predatory beetles and spiders are among the most important terrestrial predators that feed on larvae and pupae (Hill, 2008).

Insect pathogens (such as the insect fungi *Beauveria bassiana* and *Metarhizium anisopliae*) have also demonstrated the ability to infect the larvae of some leaf miners under moist conditions (FAO, 2018; CABI, 2023).

Fourth. Chemical control:

Chemical pesticides are used when infestation exceeds the economic threshold, and are best used in combination with other methods. Treating seeds with systemic pesticides also reduces seedling infestation in the early stages of growth (EPPO, 2022). Foliar spraying may be of limited effectiveness due to larval feeding within leaf tissues, but it can be beneficial against young larvae before tunneling is complete (Kalash & Said, 2004 ؛Hill, 2008).

Therefore, the following must be considered:

- Precise timing of spraying
- Adherence to recommended dosages
- Avoiding negative impacts on beneficial insects (CABI, 2023).

Fifth. Integrated Pest Management (IPM): This is the best long-term strategy and relies on:

• Early field monitoring of infestation

• Combining tillage with crop rotation

• Reducing reliance on pesticides

• Use chemical control only when necessary.

Implementation of IPM programs has been shown to reduce economic losses and maintain ecological balance (FAO, 2018; EPPO, 2022).

Wheat Stem Sawfly

Classification

Scientific name: *Cephus pygmaeus* (Linnaeus, 1767)

Order: Hymenoptera

Family: Cephidae

Common name: Wheat Stem Sawfly

Distribution

This species is one of the most important winter cereal pests in Europe, western Asia, and North Africa (Smith, 1989; Nansen et al., 2005).

Host Plants

The wheat sawfly infests cereal crops, primarily wheat, barley, and rye. Wheat is preferred because its thick stem provides suitable protection for the larvae during their development (Hatchett *et al*., 1987 ؛Shanower & Hoelmer, 2004).

Morphological Description

The adult insect is relatively small, measuring between 6–8 mm in length. It has a slender, glossy black body and transparent wings. The female possesses a saw-like ovipositor. Females have yellow transverse bands on abdominal segments

(4-6-9), while males have yellow bands on abdominal segments (3-4-6-9). The legs are yellow(Encyclopedia of Entomology, 2008).

Egg: Oval-shaped, translucent white, laid singly within stem tissue (Smith, 1989).

Larva: Legless, yellow head and body, with a distinctive shape. Larvae live within the infected stem and reach a length of approximately 14 mm when fully grown (Nansen et al., 2005).

Pupa: Forms within the plant stem near the soil surface, surrounded by dry plant tissue (Hatchett et al., 1987). (Figure 16)

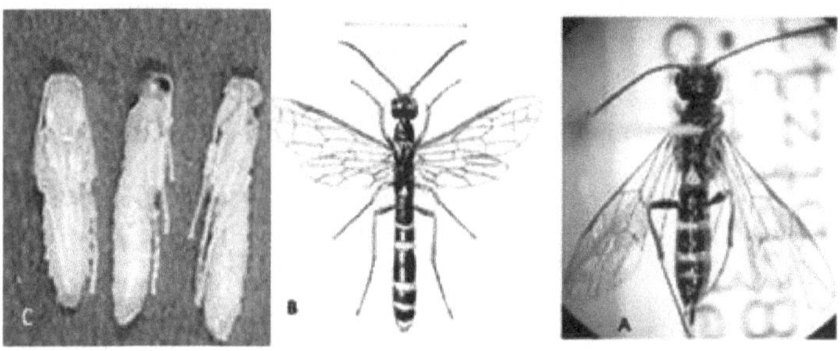

Figure (16): A. Female - B. Male - C. pupae

Damage Symptoms

Yellowing and death of the infected stem, poor grain filling, easy breakage of the stem at soil level (lodging phenomenon), significant decrease in productivity.

The primary damage results from larvae feeding inside the stem and cutting the vascular tissue, preventing the transfer of water and nutrients.

The spread of the field mouse is also observed in the infected field due to the presence of some of the fallen grains in the field, and the infected plants are characterized by the ease with which the infected ears of grain can be pulled out from them (Shanower & Hoelmer, 2004; Nansen et al., 2005). (Figure 17)

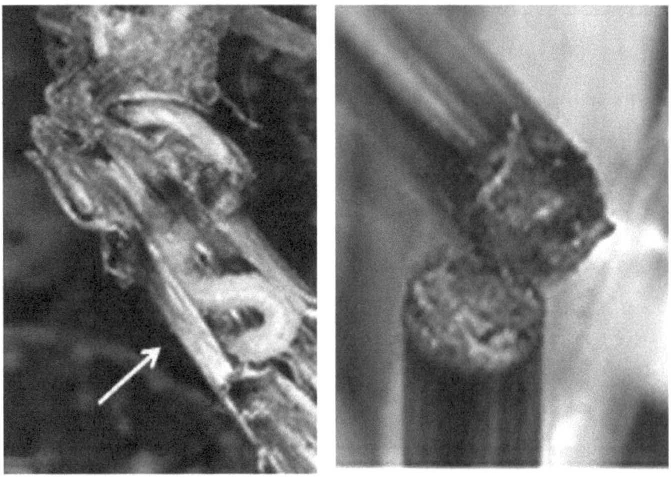

Figure (17): the damage and symptoms of infestation by the wheat sawfly.

Life Cycle

The insect overwinters as a fully grown larva inside the lower part of the wheat stalk, or the stalk remnants after harvest, where it remains in a state of physiological dormancy until the following spring (Smith, 1989; Hatchett et al., 1987).

With rising temperatures in late winter and early spring, the larva resumes its activity and pupates inside the stem tunnel, without emerging into the soil (Nansen et al., 2005). The adults emerge in late spring, coinciding with the elongation stage in wheat, where they gnaw a circular hole in the stem to exit (Shanower & Hoelmer, 2004).

After mating, the female lays eggs inside the young wheat stalk, usually one egg per stalk, using a saw-like ovipositor (Hatchett et al., 1987). The eggs hatch within a few days, and the larvae begin feeding on the inner tissue of the stalk, gradually moving downwards.

Upon reaching maturity, the larva gnaws an inner ring in the stem without cutting the outer bark, leading to weakening and eventual breakage of the stem. The insect has one generation per year (univoltine) (Smith, 1989; Nansen et al., 2005). (Figure 18)

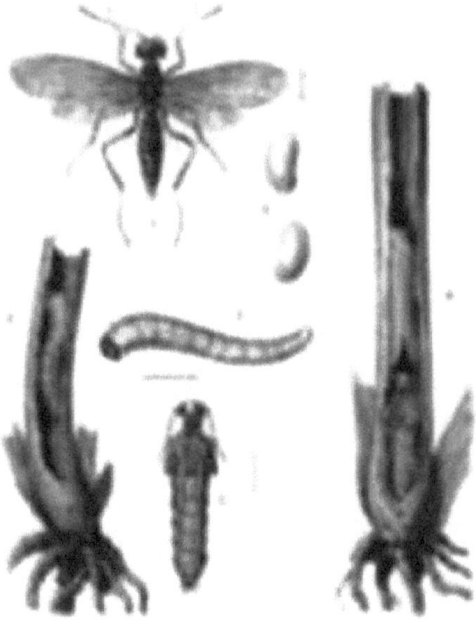

Figure (18): the life cycle stages of the saw-scaled wasp

Control methods:

Planting resistant varieties with a hardy stem, deep plowing after harvest, removal of crop residues, biological balance (natural enemies) (Shanower & Hoelmer, 2004).

Hessian fly

Scientific name: *Mayetiola destructor* (Say)

Order: Diptera

Family: Cecidomyiidae

The Hess fly is considered one of the most important wheat pests worldwide (Britannica, 2025).

The Hess fly is one of the most serious wheat pests globally, causing significant losses due to its direct impact on stalk growth and ear formation (Harris *et al*., 2003; Buntin, 1999).

Host Plants

The Hess fly infests plants of the grass family, most notably: wheat (the primary host), barley, rye, and some wild grasses. Wheat is more susceptible to infestation, especially in the early stages of growth (Hatchett *et al*., 1990; Buntin, 1999).

Morphological Description

The adult resembles a small mosquito, measuring 3–5 mm in length, with a dark brown to black color. It has long legs and transparent wings. The insect is a weak flier and has a short lifespan (Harris *et al*., 2003).

The egg is elongated and oval, initially orange-red when laid, then becoming transparent. It is laid on the upper surface of wheat leaves (Hatchett *et al*., 1990).

Larva: Legless, very small in the first instar, white then turning light green. The final instar transforms into what is known as the flaxseed stage (Buntin, 1999).

Pupa: Forms inside the hardened larval shell. It is not a clearly defined free pupa (Harris et al., 2003) (Figure 19).

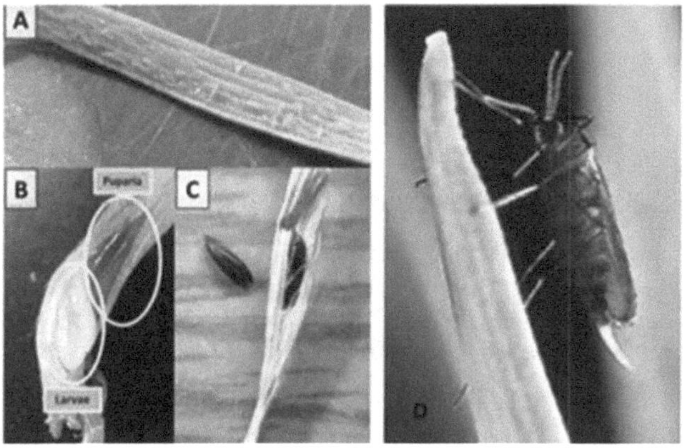

Figure (19): the insect's life cycle: A. Egg-shaped; B. Larva, pupa; C. Pupa in its final stages; D. Adult insect

Damage and Symptoms Infestation by Hess's fly causes weak or stunted seedling growth, leaf curling and shortened internodes, formation of weak or sterile plants, seedling death in early infestations, and a decrease in the number of spikes and grains (Hatchett *et al.*, 1990; Buntin, 1999). (Figure 20)

Figure (20): the symptoms of Hess's fly infestation

Life Cycle

Hess fly spends unfavorable periods of the year (extreme heat or cold) in the fully encapsulated larval stage (flaxseed stage), inside stem tissues near the base of the plant or between leaf sheaths (Harris *et al.*, 2003; Buntin, 1999).

When environmental conditions improve, the larva resumes its activity inside the hardened shell and gradually transforms into a pupa without leaving its place on the plant (Hatchett *et al.*, 1990). The adult insects then emerge in the autumn or spring depending on the planting date and climatic conditions, and live for a short period not exceeding a few days, with their main goal being reproduction (Harris *et al.*, 2003).

Immediately after mating, the female lays dozens of eggs on the upper surface of wheat leaves. The eggs hatch within 3–7 days depending on the temperature (Buntin, 1999). The newly hatched larvae crawl towards the base of the plant, where they feed on the meristematic tissue at the lower nodes.

Causing growth inhibition and stem deformation, the larva, upon reaching maturity, transforms into the flaxseed stage, which is responsible for the continued infestation from one season to the next (Hatchett *et al.*, 1990; Harris *et al.*, 2003). The insect has 2–4 generations annually, depending on climatic conditions and the agricultural region.

Control methods include: planting at recommended times, using resistant varieties, removing crop residues, and practicing crop rotation. Biological control (natural enemies) can also be employed (Buntin, 1999; Harris *et al.*, 2003).

Cereal Leaf Beetle

Scientific name: *Oulema melanopus* Linnaeus, 1758

Order: Coleoptera

Family: Chrysomelidae

According to Linnaeus' classification as cited in modern entomological sources such as Evans 2007 and EPPO 2022

Original Habitat

Studies indicate that the insect's original habitat is Central and Eastern Europe, before it later spread to Western Asia and North America via grain transport (Nelson *et al.*, 2018; EPPO, 2022).

Geographic Distribution

Today, the insect is widespread in Europe, Western Asia (including wheat and barley growing regions), and North America, particularly in temperate environments with cold winters and dry summers (according to FAO 2020 and Eppo 2022 reports).

Hosts

Main hosts: wheat, barley, oats, and occasionally wild grasses, as reported in field pest surveys (Smith 2015; FAO 2020).

Morphological description

Adult insect: The adult insect is 5–6 mm long, with metallic copper-red forewings, and glossy black head, thorax, and legs. It has an oblong shape and simple antennae. General characteristics are documented in Evans (2007) and CSIRO (2019). (Figure 21)

Larva: Resembles a soft-bodied caterpillar but is covered with a sticky layer of excrement and mucus for protection. The body is pale yellow beneath the slimy coating. It is approximately 4–5 mm long. (Description consistent with European Pest Control Reports (EPPO 2022; Smith 2015)). Eggs: Laid singly on the upper side of the leaf. The egg is yellowish-orange in color. Pupa: Forms in the soil within a small earth chamber.

Figure (21): the adult insect of the wheat leaf beetle.

Damage and Symptoms:

The larvae feed on the interveinal tissue of the plant leaf, leaving longitudinal white lines resembling peeling or scraping. This leads to reduced photosynthesis and poor grain filling, resulting in a yield decrease of up to 20% in uncontrolled fields (Nelson *et al.*, 2018; FAO 2020). Damage is more severe when the upper leaves are infested during the grain-filling stages.

Life Cycle

The adult insects overwinter in plant debris or soil. In spring, they begin feeding on young leaves and then mate. Females lay eggs on the leaves for 7–10 days until they hatch. The larvae emerge and remain for 2–3 weeks, representing the most damaging stage. After reaching maturity, they pupate in the soil for 10–14 days.

The insect has one generation per year in cold climates, and an additional generation may appear in warmer environments (EPPO 2022; Evans 2007).

Control Methods: Integrated Management

First: Agricultural Control:

This includes tilling the soil in winter to kill pupae and removing grassy weeds around the field. Also, selecting genetically robust and tolerant varieties (FAO IPM guidelines 2020).

Second: Biological control:

The parasitoid Tetrastichus julis is one of the most important biological control agents used in Europe and America (USDA 2019).

This parasitoid causes a reduction of up to 70% in larval populations.

Third: Chemical control:

Systemic and contact insecticides such as pyrethroids are used when the infestation reaches the economic threshold.

Economically, according to EPPO 2022, it is usually defined as: 1–2 larvae per leaf in the critical growth stage. (according to EPPO Pest Control Thresholds 2022).

Winter grain chafer

Scientific name: *Anisoplia* spp.

Order: Coleoptera

Family: Scarabaeidae (Medvedev, 1964; Hill, 2008).

The most important species affecting wheat in the Middle East and Mediterranean region are:

Anisoplia austriaca - *Anisoplia desertorum* (Kryzhanovsky, 1995 ؛CABI, 2023).

Origin

The winter grain crop originated in Eastern Europe and Western Asia, and then gradually spread to grain-growing regions with temperate and semi-arid climates, especially in steppe environments (a type of open natural environment). It is characterized by the spread of short and medium grasses with almost no trees, and is a transitional environment between desert areas and areas with dense vegetation cover. (Kryzhanovsky, 1995; FAO, 2017).

Distribution

Anisoplia species are found throughout the world, including the Middle East (Syria, Iraq, Turkey, Iran), North Africa, Central and Eastern Europe, and Central Asia. They are common pests in poorly managed agricultural wheat fields (CABI, 2023; FAO, 2017).

Host plants

The winter grain beetle infests crops of the grass family, most notably: wheat, barley, oats, and some wild grasses (Medvedev, 1964; Hill, 2008).

Morphological description

Adult insect: A medium-sized beetle, 10–15 mm long, yellowish-brown to black with a slight sheen, elytra often with dark spots, mouthparts strong and chewing. (Figure 22) Larva: White, C-shaped, brown head, prominent thoracic legs, lives in the soil (Medvedev, 1964; Kalash, 2007).

Figure (22): the adult insect of the winter grain-making plant – and symptoms of infestation

Damage and symptoms

The adult insect feeds on mature ears of grain, gnawing at the milky and doughy stages, resulting in a significant reduction in grain weight and yield.

The larvae, on the other hand, feed on wheat roots, causing plant weakness and yellowing. In severe infestations, this can lead to seedling death (Hill, 2008; FAO, 2017). (Figure 22)

Life cycle

The winter grain weevil is a holometabolous insect, and its life cycle typically lasts one year, though it can extend to two years in some environments. After mating in late spring and early summer, the female lays her eggs in the soil at a depth of 5–15 cm in or near wheat fields. The eggs are spherical to oval and white. The incubation period lasts from 10 to 20 days, depending on temperature and soil moisture (Medvedev, 1964; Kalash, 2007).

The larvae (the main damaging stage) hatch from the eggs and feed directly on the roots. The larva undergoes three larval instars, with this being the longest stage in the life cycle. The larvae spend the summer and autumn in the soil.

They then enter a state of overwintering at greater depths as temperatures decrease. The larval stage lasts between 8 and 10 months, and may extend to more than a year in some species (Hill, 2008; FAO, 2017). The following spring, the fully grown larva pupates inside an earthen chamber. The pupa is free-form, yellowish-white, and this stage lasts from 2 to 4 weeks (Medvedev, 1964; Kryzhanovsky, 1995). The adult insects appear in late spring and early summer, feed on wheat ears, then mate and begin laying eggs, thus starting a new life cycle. The adult insect has a relatively short lifespan compared to the larva (CABI, 2023; Hill, 2008).

Control methods

First: Agricultural control:

Deep plowing to break up larval development sites, crop rotation, and removal of grassy weeds.

Second: Biological control:

Encouraging natural predators and using insect pathogens.

Third: Chemical control:

Using soil-borne herbicides when necessary, and applying treatments in a timely manner before the adult insect reaches the ears (FAO, 2017 ؛CABI, 2023).

Wheat and Grass Thrips

Wheat and cereals are attacked by several types of thrips, the most important of which are:

Wheat thrips (*Haplothrips tritici*, Kurdjumov, 1912)

Cereal thrips (*Limothrips cerealium*, Haliday, 1836)

Limothrips denticornis (Haliday, 1836)

Classification

Order: Thysanoptera Family: Thripidae and Phlaeothripidae Thrips insects are important secondary pests of wheat and may become economic pests when climatic conditions are favorable, especially in arid and semi-arid regions (Lewis, 1997; Mound & Kibby, 1998).

Host Plants

Thrips infest a wide variety of grasses, including wheat, barley, oats, corn, and wild grasses. Wheat is a preferred host, especially during the heading and flowering stages (Kirk & Terry, 2003; Lewis, 1997).

Morphology

The adult insect: very small (1–2 mm), slender elongated body, dark brown to black in color, narrow wings surrounded by long hairs, piercing-sucking mouthparts (Mound & Kibby, 1998). (Figure 23)

The egg: elongated oval, transparent when laid, laid inside plant tissues (leaves or spikes) (Lewis, 1997).

Nymphs: Resembling the adult insect but wingless, light yellow in color, fast-moving, and actively feeding on plant tissue (Kirk & Terry, 2003). Prepupal and **pupa stages:** Relatively dormant stages that do not feed, often found in the soil or among plant debris (Lewis, 1997).

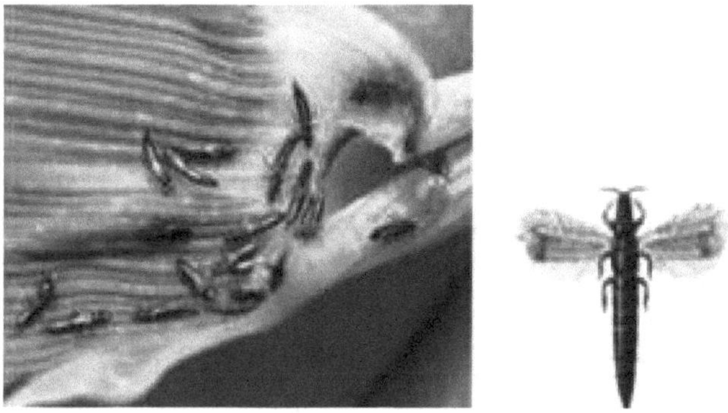

Figure (23): the adult thrips insect.

Damage and Symptoms

The damage results from the puncture of plant cells and the absorption of their contents by both the larva and the adult insect. Symptoms appear on wheat and grasses as follows: silver or yellow spots on the leaves, curling and drying of leaf tips, deformed ears and poor grain filling, and a decrease in grain weight. Severe infestation also causes stunting of plants.

Some types of thrips may also contribute to the transmission of viral pathogens in other crops, although this is limited in wheat (Kirk & Terry, 2003; Lewis, 1997).

Life Cycle

Thrips typically overwinter as adults or pupae within plant debris, topsoil, or under the leaf sheaths of grasses (Mound & Kibby, 1998; Lewis, 1997). With improved conditions in spring, the females become active and lay eggs within leaf tissue or spikelets, protecting them from environmental factors and natural predators (Kirk & Terry, 2003).

The eggs hatch within several days, and the nymphs begin to feed intensively on the leaves or parts of the spike; this stage is the most damaging to the plant (Lewis, 1997). After the nymphs have completed their development, they move to the soil or plant debris to enter the prepupal stage and then the pupal stage (Mound & Kibby, 1998).

The adult insects emerge to restart the life cycle, and their reproduction continues as long as suitable conditions of heat and relative dryness are present. The number of generations increases in warm and dry regions (Kirk & Terry, 2003). (Figure 24)

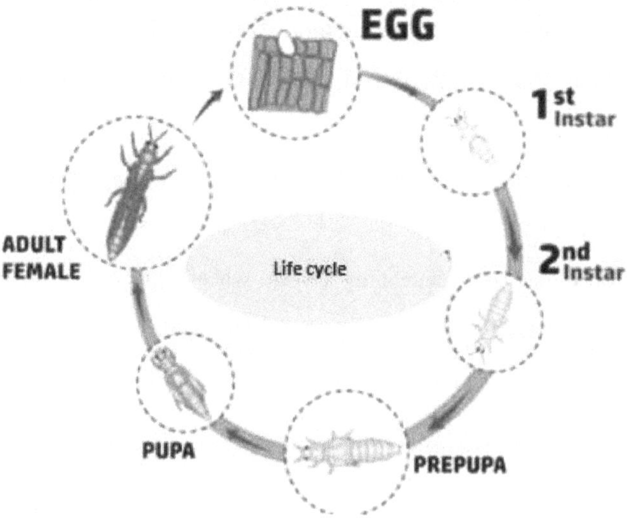

Figure (24): the stages of the life cycle of the thrips insect.

The main factors contributing to increased thrips infestation are:

- High temperatures
- Drought and low rainfall
- High plant density
- Neglect in crop residue management (Lewis, 1997; Kirk & Terry, 2003)

Control methods

Planting at appropriate times, regular irrigation to reduce drought, removal of grassy weeds, encouragement of beneficial insects, use of pesticides only when necessary (Mound & Kibby, 1998).

The concept of the economic threshold in wheat

The economic threshold (ET) is defined as the level of pest density at which control action must be taken to prevent damage from reaching the level of economic loss, i.e., when the cost of control equals the value of the expected crop loss (Pedigo & Rice, 2014).

These thresholds vary depending on the wheat's growth stage, the variety planted, environmental conditions, crop price, and control costs (Dent, 2000). The following are the economic thresholds for each of the aforementioned insects:

First: *Eurygaster integriceps*

The economic threshold for the sunn pest depends more on the percentage of damaged grains and the percentage of white grains than on the number of insects alone, because qualitative damage to wheat grains directly affects the quality of flour and gluten (Hariri *et al.*, 2000).

Studies in Turkey and Iraq have shown that exceeding a certain limit of damage leads to a significant decrease in the market value of grains, making intervention necessary before the milky maturity stage when nymph density reaches critical levels calculated locally (Mutlu et al., 2016).

Secondly: wheat aphids (*Rhopalosiphum padi-, Sitobion avenae*)

Aphids are piercing-sucking pests with rapid reproduction. Their danger lies in sucking plant sap and transmitting viruses. Guidelines indicate that the economic threshold is usually achieved when the aphid population reaches about 12–15 individuals per stem before the milk stage.

This is where the noticeable effect on grain weight and count begins (Blackman & Eastop, 2000; Pedigo & Rice, 2014).

Third: The rural armyworm, *Spodoptera frugiperda*

It is a polyphagous chewing pest. Studies indicate that the economic threshold in cereal crops is approximately reached when the larval population reaches about 10 larvae per square meter, at which point leaf damage becomes significant in photosynthetic (FAO, 2018).

Fourth: Cereal seedling moth, *Zabrus tenebrioides*

The primary damage occurs in the seedling stage as a result of adults and larvae feeding on young leaves. European studies have indicated that the economic threshold may be reached at 2–3 individuals per square meter, especially in early plantings, where the loss of plant density significantly impacts the final yield (Holland, 2002).

Fifth: Cutworms, *Agrotis* spp.

Cutworms cause significant losses by cutting seedlings at the soil surface. References indicate that the economic threshold is often reached when there are several larvae per square meter, especially in the early stages of wheat growth, as the plant cannot compensate for the loss (Hill, 1987; Dent, 2000).

Sixth: Cereal Leafminer, *Syringopais temperatella*

The economic threshold for this insect is related to the percentage of infested plants and the severity of leaf curling and drying. Studies in the Middle East have shown that control intervention becomes economically justified when infestation exceeds approximately 20%.

From plants in the branching stage, cumulative damage leads to a decrease in effective leaf area and consequently productivity (Shojai *et al.*, 2010; EPPO, 2019).

Seventh: Wheat sawfly, *Cephus pygmaeus*

This insect is a pest for which it is difficult to establish a precise numerical economic threshold due to the nature of the internal stem damage. Therefore, integrated management relies on indirect indicators such as lodging percentage or market breakage. Studies recommend using hard-stemmed varieties when infestation exceeds limits that affect plant stability and yield (Beres *et al.*, 2011).

Eighth: Hess fly, *Mayetiola destructor*

There is no clear economic threshold for infestation within the growing season for this insect, as damage occurs early within plant tissues. Therefore, management relies on preventative measures such as selecting resistant varieties and adjusting planting dates, as confirmed by most classic plant protection references (Hatchett *et al.*, 1987).

Ninth: Cereal leaf beetle, *Oulema melanopus*

The larvae cause tissue scraping of the leaves, reducing the green area. The economic threshold is usually reached when there is approximately one larva per plant in the elongation stage, where the damage leads to a significant decrease in yield (Buntin *et al.*, 2004).

Tenth: Winter grain, *Anisoplia* spp.

The economic threshold for this insect is determined during the flowering period, when the females lay their eggs on the ears. A density of one female per 4–5 ears is considered an economic indicator for making a control decision, because the larvae feed directly on the developing grains, causing a decrease in weight and quality (Oakley *et al.*, 1998).

Eleventh: Wheat Thrips (*Thrips* spp.)

Thrips cause leaf and spike deformation and reduced grain fullness. Extension guidelines state that the economic threshold is often reached when the population reaches 7–8 individuals per stalk in the early stages before spike emergence (Lewis, 1997).

Table (1): containing the scientific and common names of insect pests that infest wheat

Common name	Scientific name
Sunn pest	*Eurygaster integriceps* Puton
Bird cherry–oat aphid	*Rhopalosiphum padi (*L.*)*
Fall armyworm	*Spodoptera frugiperda* (J.E.Smith,1797)
Cereal ground beetle	*Zabrus tenebrioides* (Goeze)
Cutworms	*Agrotis* spp.
Cereal leaf miner	*Syringopais temperatella* (Lederer)
Wheat Stem Sawfly	*Cephus pygmaeus* (Linnaeus, 1767)
Hessian fly	*Mayetiola destructor* (Say)
Cereal leaf beetle	*Oulema melanopus* (L.)
Cereal chafer beetles	*Anisoplia spp.*
Wheat thrips	*Haplothrips tritici* L.

Table (2): containing the scientific and common names of the most important crops mentioned in the book.

Common name	**Scientific name**
Wheat	*Triticum aestivum* L
Barley	.*Hordeum vulgare* L
Oat	.*Avena sativa* L
Rye	*Secale cereale* L
Maize / Corn	*Zea mays* L

References

1. Anderson, J. W., Baird, P., Davis, R. H., Ferreri, S., Knudtson, M., Koraym, A., Waters, V., & Williams, C. L. (2000). Whole grain foods and cardiovascular disease risk. Journal of Nutrition, 130(6), 2037–2042. https://doi.org/10.1093/jn/130.6.2037
2. Ahdb. (2023). Basic biology of the bird cherry-oat aphid (*Rhopalosiphum padi*). Retrieved from https://ahdb.org.uk/knowledge-library/basic-biology-of-the-bird-cherry-oat-aphid
3. Agrobiotop. (2023). English grain aphid (*Sitobion avenae*). Retrieved from https://www.agrobiotop.com/en/english-grain-aphid-sitobion-avenae
4. Al-Zyoud, F., & Katbeh-Bader, A. (2011).Biology and seasonal activity of *Syringopais temperatella* on cereal crops in Jordan.Jordan Journal of Agricultural Sciences, 7(2), 217–228.
5. Arab Journal of Plant Protection. (2023).
6. ARC ,Egypt. (2023). Leafhopper Transmission Studies.
7. Arab Journal of Plant Protection. (2022).
8. APS Plant Health Journal. (2021).
9. Backlund, H. O. (1986).Studies on the biology, ecology, and damage of *Zabrus tenebrioides* (Coleoptera: Carabidae.(Zeitschrift für Angewandte Entomologie, 101, 241–251.
10. Brown, E. S. (1962). The biology and ecology of the Sunn pest, *Eurygaster integriceps* Put. Bulletin of Entomological Research, 53, 417–438.
11. Britannica. (2025). Hessian fly. Encyclopedia Britannica. Retrieved from: https://www.britannica.com/animal/Hessian-fly

12. Buntin, G. D. (1999). Hessian fly (Diptera: Cecidomyiidae) biology and management. Annual Review of Entomology, 44, 429–446.
13. Buntin, G. D., Flanders,K.L., Slaughter,R,W,.& Delamar,Z.D. (2004). Damge loss assessment and control of ceeeal leaf beetle (Coleoptera: Chrysomellidae) in winter wheat. *Journal of Economic Entomo*logy, 97(2),374-382. Httos//doi.org/10.1093/jee/97.2.374.
14. Blackman, R. L., & Eastop, V. F. (2000). Aphids on the World's Crops. John Wiley & Sons.
15. Beres, B. L., Cárcamo, H. A., & Weaver, D. K. (2011). Integrated management of wheat stem sawfly. Canadian Journal of Plant Science.
16. CIMMYT. (International Maize and Wheat Improvement Center) (2021). Aphid Transmission of Cereal Viruses..
17. CIMMYT (International Maize and Wheat Improvement Center). (2022). Wheat Aphid Dynamics.
18. CABI (2023).*Anisoplia* spp Crop Protection Compendium.CAB International.
19. CABI (Centre for Agriculture and Bioscience International). (2023).*Spodoptera frugiperda* (fall armyworm) – Datasheet.In: Invasive Species Compendium.Wallingford, United Kingdom: CABI.
20. CABI (2022).(*Agrotis* spp. – Datasheet.Centre for Agriculture and Bioscience International, UK.
21. Capinera, J. L. (2017).Fall armyworm, *Spodoptera frugiperda* (J.E. Smith) (Insecta: Lepidoptera: Noctuidae.(University of Florida, IFAS Extension.Gainesville, Florida, USA.

22. Capinera, J. L. (2018).Cutworms and armyworms (Lepidoptera: Noctuidae).In: Encyclopedia of Entomology.Dordrecht, Netherlands: Springer.
23. Capinera, J. L. (Editor). (2008). Encyclopedia of Entomology (2nd ed.). Springer Science and Business Media.
24. Capinera, J. L. (2017). Handbook of Vegetable Pests 2nd ed.(.Academic Press, Elsevier, USA).Cutworms section.
25. CSIRO (Commonwealth Scientific and Industrial Research Organisation). (2019). Insects of agricultural importance.CSIRO Publishing.Canberra, Australia.
26. CABI (Centre for Agriculture and Bioscience International). (2022).*Zabrus tenebrioides* (Goeze, 1777) – Datasheet.In: Invasive Species Compendium.Wallingford, UK: CABI.
27. CABI (2022). *Eurygaster integriceps* (Sunn pest). In: Invasive Species Compendium. CAB International.
28. CABI (2023).*Syringopais temperatella* – Crop Protection Compendium.CAB International.
29. CABI (Centre for Agriculture and Bioscience International). (2022).*Agrotis* spp. (Cutworms) Datasheet.In: Crop Protection Compendium.Wallingford, United Kingdom: CABI.
30. CABI (Centre for Agriculture and Bioscience International). (2022).*Syringopais temperatella* (Lepidoptera: Scythrididae) – Datasheet. In: Crop Protection Compendium.Wallingford, United Kingdom: CABI.

31. Chiavaroli, L., Mirrahimi, A., Sievenpiper, J. L., Jenkins, D. J., & Augustin, L. S. A. (2018). Whole grains and glycemic control in diabetes. Nutrients, 10(6), 649. https://doi.org/10.3390/nu10060649
32. Davis, T. *et al*. (2020). Wheat Stem Sawfly Review.
33. Day, R., Abrahams, P., Bateman, M. (2017). Fall armyworm: Impacts and implications. Outlooks on Pest Management, 28(5), 196–201.
34. Dent, D. (2000). Insect Pest Management. CABI Publishing.
35. Encyclop'Aphid. (2023). *Sitobion avenae* biology and damage. Retrieved from https://eng-encyclopedie-pucerons.hub.inrae.fr/species/aphids/sitobion/s.-avenae
36. EPPO (2013). Eurygaster integriceps. EPPO Bulletin, European and Mediterranean Plant Protection Organization.
37. EPPO (European and Mediterranean Plant Protection Organization). (2021).*Spodoptera frugiperda*.EPPO Global Database.Paris, France: EPPO Secretariat.
38. EPPO (2021).*Agrotis* spp. – Biology and distribution. European and Mediterranean Plant Protection Organization.
39. EPPO (European and Mediterranean Plant Protection Organization). (2021). *Syringopais temperatella*. EPPO Global Database. Paris, France: EPPO Secretariat.
40. EPPO (2022). *Syringopais temperatella* data sheet.European and Mediterranean Plant Protection Organization.
41. EPPO (2025). *Cephus pygmeus* – Taxonomic classification. EPPO Global Database.
42. EPPO (European and Mediterranean Plant Protection Organization). (2022).*Oulema melanopus* (L.) – Cereal leaf beetle: EPPO datasheet on pests

recommended for regulation.EPPO Global Database.Paris, France: EPPO Secretariat.

43. EPPO (European and Mediterranean Plant Protection Organization). (2022).EPPO Pest Control Thresholds: Cereals pests.EPPO Standards and Guidelines, Series PP1 (Good Plant Protection Practice.(Paris, France.
44. EPPO (2019). *Syringopais temperatella* datasheet. EPPO Bulletin.
45. EPPO (European and Mediterranean Plant Protection Organization). (2021).*Agrotis ipsilon* and *Agrotis segetum*.EPPO Global Database.Paris, France: EPPO Secretariat.
46. EPPO (European and Mediterranean Plant Protection Organization). (2021).*Zabrus tenebrioides*.
47. EPPO Global Database.Paris, France: EPPO Secretariat.
48. EPPO. (2022). *Spodoptera frugiperda*. EPPO Global Database.
49. Evans, A. V. (2007).Field guide to insects and spiders of North America.New York, USA: Sterling Publishing Co., Inc.
50. El-Lakwah, F. A., El-Sheikh, E. A., & Abdel-Rahman, R. S. (2019).Economic damage and population dynamics of cutworms (*Agrotis* spp.) infesting cereal crops.Egyptian Journal of Agricultural Research, 97(2), 421–435.
51. FAO (Food and Agriculture Organization of the United Nations). (2022).Fall armyworm (*Spodoptera frugiperda*): Global action for monitoring, management and prevention.FAO Plant Production and Protection Division.Rome, Italy.
52. FAO (2018). Integrated management of sunn pest in wheat. Food and Agriculture Organization of the United Nations, Rome.
53. FAO (2018).Insect pests of cereals in Near East region.Food and Agriculture Organization of the United Nations.

54. FAO (2018). Fall Armyworm Monitoring and Management. FAO Plant Production Series.
55. FAO (2017).Cereal pests in the Near East region. Food and Agriculture Organization of the United Nations.
56. FAO (Food and Agriculture Organization of the United Nations). (2023). Major insect pests of wheat and barley and their management.FAO Plant Production and Protection Division.Rome, Italy.
57. FAO (Food and Agriculture Organization of the United Nations). (2020).Integrated Pest Management (IPM) guidelines for cereal crops.FAO Plant Production and Protection Division.Rome, Italy.
58. FAO. (2018). Integrated management of the Fall Armyworm. Food and Agriculture Organization of the United Nations.
59. FAO (Food and Agriculture Organization of the United Nations). (2020).Cereal crop pest management in temperate zones.FAO Plant Production and Protection Paper.Rome, Italy.
60. FAO (Food and Agriculture Organization of the United Nations). (2020).Cereal pests in dry and semi-arid regions of the Near East.FAO Plant Production and Protection Division.Rome, Italy.
61. FAO. (2025a). Short-term outlook for global cereal supply and demand 2025/26. Food and Agriculture Organization of the United Nations. https://www.fao.org/worldfoodsituation/csdb/ar
62. FAO. (2025b). Global cereal production trends and forecasts for 2025. Food and Agriculture Organization of the United Nations. https://gate.ahram.org.eg/News/5279856.aspx
63. FAO (2011). Wheat and health. Food and Agriculture Organization of the United Nations.

64. FAO. (2022). Wheat Pest Management in West Asia.
65. FAO. (2020). Grain Storage Manual.
66. FAO , 2022. Egypt Country Report.
67. FAO Early Warning. (2023). Fall Armyworm Situation Update.
68. FAO (2017). Wheat production and sustainable agriculture. Food and Agriculture Organization of the United Nations.
69. FAO GIEWS. (2025). Country Brief: Syrian Arab Republic — Wheat import requirements and production estimates 2025/26. https://www.fao.org/giews/countrybrief/country.jsp?code=SYR&lang=ar
70. Goergen, G., Kumar, P. L., Sankung, S. B., Togola, A., & Tamò, M. (2016).First report of outbreaks of the fall armyworm *Spodoptera frugiperda* (J. E. Smith) (Lepidoptera, Noctuidae), a new alien invasive pest in West and Central Africa.PLOS ONE, 11(10), e0165632.https://doi.org/10.1371/journal.pone.0165632
71. Hatchett, J. H., Martin, T. J., & Jones, E. T. (1987). Biology and ecology of the wheat stem sawfly, *Cephus pygmaeus*. Annals of the Entomological Society of America, 80(3), 349–354.
72. Hatchett, J. H., Starks, K.J., & Webster,J.A. (1987). Insect and mite pests of wheat . In E.G.Heyne(Ed), Wheat and Wheat Improvement (2nd ed., Agronomy Monograph No. 13, pp. 6625 -675) . American Society of Agronomy, Madison, WI,USA.
73. Holland, J. M. (2002). The agroecology of *Zabrus* species. Agriculture, Ecosystems & Environment.

74. Harrison, R. D., Thierfelder, C., Baudron, F., *et al*. (2019). Agro-ecological management of fall armyworm. *Journal of Integrated Pest Management*, 10(1).

75. Hariri, G., Williams, P. C., & El-Haramein, F. J. (2000). Influence of sunn pest damage on wheat quality. Journal of Cereal Science, 32, 137–145.

76. Hill, D. S. (1987). Agricultural Insect Pests of Temperate Regions. Cambridge University Press.

77. Hill, D. S. (2012).Agricultural Insect Pests of Temperate Regions and Their Control.Cambridge University Press, UK.

78. Hill, D. S. (2008).Pests of Crops in Warmer Climates and Their ControlSpringer Science & Business Media.

79. Hill, D. S. (2008).Pests of Crops in Warmer Climates and Their Control. Springer Science & Business Media.

80. Harris, M. O., Rose, S., & Flanders, K. L. (2003). The Hessian fly: biology, genetics, and management. Entomological Society of America, Lanham, Maryland.

81. Hatchett, J. H., Gallun, R. L., & Jennings, P. R. (1990). Hessian fly biology and resistance in wheat. *Journal of Economic Entomology*, 83(2), 395–401.

82. IPPC. (2022). Global Pest Alerts Database.

83. ICARDA. (2020). Wheat Pest Complex in Dry Regions.

84. Jenkins, D. J. A., Kendall, C. W. C., Marchie, A., Faulkner, D. A., Wong, J. M. W., de Souza, R., Emam, A., Parker, T. L., & Vidgen, E. (2002). Effect of high-fiber diets on cardiovascular disease. American Journal of Clinical Nutrition, 76(1), 5–10. https://doi.org/10.1093/ajcn/76.1.5

85. Jones, J.M. (2014). Fiber and prebiotics: mechanisms and health benefits. Nutrients, 6(6), 2117–2136.

86. JPPP. (2023). Aphid species density in wheat fields. Retrieved from https://jppp.*journals.ekb.eg/article*_420333.html

87. Kenis, M., du Plessis, H., Van den Berg, J., *et al.* (2019). Managing fall armyworm. Annual Review of Entomology, 64, 19–36.

88. Kriticos, D. J., (2017). Climatic suitability and pest risk assessment of sunn pest. Agricultural Systems, 157, 123–134.

89. Kalash, S. H. (2007). Insect pests of cereal crops in the Middle East. *Arab Journal of Plant Protection*, 25(2), 85–96.

90. Kalash, S. H., & Said, F. A. (2004).Studies on the cereal leaf miner *Syringopais temperatella* in wheat fields.Arab Journal of Plant Protection, 22(1), 45–52.

91. Kryzhanovsky, O. L. (1995). Coleoptera of Eastern Europe.St. Petersburg University Press.

92. Kfir, R. (1983). Biology, ecology and economic importance of the wheat seedling pest *Syringopais temperatella* (Lepidoptera: Scythrididae) in the Middle East. Phytoparasitica, 11(3–4), 191–200.

93. Kirk, W. D. J., & Terry, L. I. (2003). The spread of thrips. Agricultural and Forest Entomology, 5(3), 141–150.

94. Lewis, T. (1997). Thrips as crop pests. CAB International, Wallingford, UK.

95. Ludwig, D.S. (2002). The glycemic index: physiological mechanisms relating to obesity, diabetes, and cardiovascular disease. JAMA, 287(18), 2414–2423.

96. Medvedev, S. I. (1964).Scarabaeidae of the USSR. Israel Program for Scientific Translations.

97. Mound, L. A., & Kibby, G. (1998). Thysanoptera: An identification guide. CAB International, Wallingford, UK.

98. Mutlu, Ç., et al. (2016). Economic threshold of Eurygaster integriceps. *Turkish Journal of Agriculture*.
99. Nansen, C., Payton, M. E., & Runyon, J. B. (2005). Population dynamics and life history of wheat stem sawfly. Agricultural and Forest Entomology, 7(2), 127–136.
100. Nelson, D., Giles, K. L., Royer, T. A., & Elliott, N. C. (2018).Biology, ecology, and management of the cereal leaf beetle (*Oulema melanopus* L.) *Journal of Integrated Pest Management*, 9(1), 1–8.https://doi.org/10.1093/jipm/pmy005
101. Purdue University. (2023). Small grains aphids monitoring and IPM. Retrieved from https://ag.purdue.edu/department/entm/extension/field-crops-ipm/small-grains/aphids.html
102. PubMed. (2020). Aphid pesticide resistance study. Retrieved from https://pubmed.ncbi.nlm.nih.gov/33172670/
103. Prasanna, B. M., Huesing, J. E., Eddy, R., & Peschke, V. M. (2018). Fall armyworm in Africa: A guide for integrated pest management. CIMMYT.
104. Paulian, F., & Popov, C. (1980). Sunn pest or cereal bug. In: Wheat pests and their control. FAO Plant Production and Protection Paper.
105. Pedigo, L. P., & Rice, M. E. (2014).Entomology and Pest Management 6th ed..Pearson Education, USA.
106. Schuh, R. T., & Slater, J. A. (1995). True Bugs of the World (Hemiptera: Heteroptera). Cornell University Press.
107. Shanower, T. G., & Hoelmer, K. A. (2004). Biological control of wheat stem sawflies. CABI Publishing, Wallingford, UK.

108. Shafaq News. (2025). Syria's cultivated wheat area falls 13% in 2024–2025 season. https://shafaq.com/en/Middle-East/Cultivated-wheat-area-in-Syria-falls-13-in-2024-2025-season
109. Shewry, P.R., & Hey, S.J. (2015). The contribution of wheat to human diet and health. Food and Energy Security, 4(3), 178–202.
110. Smith, R. C. (1989). Wheat stem sawflies (Hymenoptera: Cephidae): Biology, ecology, and management. *Journal of Economic Entomology*, 82(5), 1231–1237.
111. Smith, J. L. (2015).Cereal pests and their management.Wallingford, UK: CABI Publishing
112. Slavin, J. (2003). Why whole grains are protective: biological mechanisms. Proceedings of the Nutrition Society, 62(1), 129–134.
113. Shewry, P.R. (2009). Wheat. Journal of Experimental Botany, 60(6), 1537–1553.
114. USDA (United States Department of Agriculture). (2019). Biological control of cereal leaf beetle (Oulema melanopus) in wheat and barley.USDA Agricultural Research Service (ARS).Washington, D.C., USA.
115. USDA. (2021). Hessian Fly Biology and Control.
116. UC IPM. (2023). Bird cherry-oat aphid on small grains. Retrieved from https://ipm.ucanr.edu/agriculture/small-grains/bird-cherry-oat-aphid.
117. USDA (United States Department of Agriculture). (2022).Cutworms and armyworms: Biology and management.USDA Agricultural Research Service (ARS).Washington, D.C., USA.
118. USDA FA0. (2025a). Egypt wheat area, production, and yield data 2025/26. https://ipad.fas.usda.gov/countrysummary/Default.aspx?crop=Wheat&id=EG

119. USDA FAO. (2025b). Egypt wheat production forecast 2025. https://ipad.fas.usda.gov/countrysummary/Default.aspx?crop=Wheat&id=EG
120. Wang, S., Wang, L., Li, D., Zhang, X., & Zhang, X. (2009). Antioxidant capacity and phenolic content of whole wheat products. Food Chemistry, 116(4), 954–960. https://doi.org/10.1016/j.foodchem.2009.03.036
121. Wikipedia. (2025a). Sunn pest. https://en.wikipedia.org/wiki/Sunn_pest
122. Wikipedia. (2025b). Angoumois grain moth. https://en.wikipedia.org/wiki/Angoumois_grain_moth
123. Wikipedia. (2023). *Sitobion avenae*. Retrieved from https://en.wikipedia.org/wiki/Sitobion_avenae
124. Xu, B., & Chang, S.K.C. (2007). A comparative study on phenolic profiles and antioxidant activities of legumes. Journal of Agricultural and Food Chemistry, 55(21), 8845–8852.
125. Yuksel, E., Imren, M.,&Canhilal, R.(2024). Potential biopesticides for the control of wheat grain beetle, *Anisoplia austriaca*(Herbst) (Coleoptera: Scarabaeidae: Rutelidae): Spinosad and entomopathogenic nematodes. Erciyes Tarim veHayvan Bilimleri Dergisi,7(1),ethabd.1441005.
126. ARC, Egypt. 2021126. Wheat Pests in Egyptian Soils.
127. 127. Syrian General Commission for Agricultural Research (SARC), 2022. Entomology Department.
128. Plant Protection Research Institute, (2022). Egypt....128
129. Syrian Ministry of Agriculture. (2023). Annual Agricultural Pests Report.
130. Egyptian Ministry of Agriculture. (2023). Agricultural Pest Management

About Authors

Prof. Dr. Mohamed Abdel-Raheem Ali Abdel-Raheem

Professor of Entomology (Biological Control)

Pests & Plant Protection Department

Agricultural and Biological Research Institute

National Research Centre.

Prof. Dr. Mohamed Abdel-Raheem Ali Abdel-Raheem, Prof. of Entomology (Biological Control), Pests & Plant Protection Department, Agricultural and Biological Research Institute, National Research Centre, Cairo, Egypt. Published publications (322) papers (87), Books and Book chapters (235), Edited books (8), (Scopus) h-index (12), Citations (339), (Google Scholar) h-index (21) , Citations (1207), (Web of Science) Publications (29), h-index (11), Citations (318), Research Gate, h-index (15), Citations (737), Reviewers in International Journals (159), Reviewed articles in Biological Control (1087), Editor in Chief in Journal (7), Assistant Editor (1), Associated Editor in Chief in Journal (7), Editorial board in Journals (44), Advisory board (1), conference Attended (27), workshop Attended (313), Symposium Attended (179), Forum Attended (4), Prize (4), Attended Others (30), Projects As PI and member (20), Training courses For Agric. Engineering (61), Training courses For Students of University (12), attending Trainees courses (19), TV, Radio & Newspaper (39), Supervisor on Ms.C. thesis (3), Supervisor on Ph.D. thesis (1), Committee Ph.D. Thesis in Egypt (2), and member in scientific Foundation (14), Giving International lectures (4), Promotion Committee Prof. international (2), Committee Ph.D. thesis international (1).

National Research Centre. 33rd ElBohouth St., Dokki, Cairo, Egypt.

Email: abdelraheem_nrc@hotmail.com, ma.abdel-raheem@nrc.sci.eg, abdelraheem_nrc@yahoo.com, profmohamedabdelraheem@gmail.com

Mobile Phone: (+2) 01155527583 – (+2) 01009580797

Scopus Author ID : 55220661700

https://www.scopus.com/authid/detail.uri?authorId=55220661700

http://scholar.google.com/citations?hl=en&user=gBtssEgAAAAJ

https://orcid.org/my-orcid?orcid=0000-0001-9240-064X

https://www.webofscience.com/wos/author/searchAbdel-Raheem, Mohamed Abdel-Raheem Ali - Web of Science Core Collection

https://www.researchgate.net/profile/Mohamed_Abdel-Raheem4/publications?sorting=recentlyAdded

http://livedna.org/20.16018

http://www.resarcherid.com/rid/R-6264-2017

https://www.webofscience.com/wos/author/record/R-6264-2017

https://www.facebook.com/share/p/1Piu6xyFmN/

- TV, Radio & Newspaper links

1. https://www.youtube.com/attribution_link?a=NsH2xGv9mY8&u=%2Fwatch%3Fv%3Dsb46rBG8-Mc%26feature%3Dshare
2. https://youtu.be/Ub8djW0tggo
3. https://m.facebook.com/story.php?story_fbid=492907069288338&id=100076011247165
4. https://youtu.be/dOujUAZmTdw
5. https://m.facebook.com/story.php?story_fbid=492907069288338&id=100076011247165
6. https://fb.watch/gPGrDuzZFX/
7. https://www.youtube.com/watch?si=4UfyQaeHVLPiVR5J&v=nXV4NT-no7o&feature=youtu.be
8. https://www.youtube.com/watch?v=sRH44W6TbcQ
9. https://youtu.be/uq9Qj3AEBHk?si=ymaX06huG2AXJESu
10. https://www.facebook.com/share/v/Qg5s6HW3hD8PLBzp/?mibextid=xfxF2i
11. https://el-yom.com/news/archives/1092767
12. https://el-yomnews.com/?p=1156515

13. https://el-yomnews.com/%D8%A7%D9%84%D8%AF%D9%83%D8%AA%D9%88%D8%B1-%D8%B9%D8%A8%D8%AF%D8%A7%D9%84%D8%B1%D8%AD%D9%8A%D9%85-%D9%85%D8%AD%D9%85%D8%AF-%D8%B9%D9%84%D9%8A-%D9%8A%D8%AA%D8%AD%D8%AF%D8%AB-%D9%84-%D9%84%D9%84/
14. https://el-yomnews.com/?p=1179515
15. https://www.facebook.com/share/v/4MM85wKRPxsYLqTr/
16. http://https//el-yomnews.com/%D8%A7%D9%84%D8%AD%D8%B4%D8%B1%D8%A7%D8%AA-%D9%81%D9%8A-%D8%AE%D8%AF%D9%85%D8%A9-%D8%A7%D9%84%D8%B9%D8%AF%D8%A7%D9%84%D8%A9-%D9%85%D8%AD%D8%A7%D8%B6%D8%B1%D8%A9-%D8%AA%D9%83%D8%B4%D9%81-%D8%AF%D9%88/
17. https://youtube.com/watch?v=F21ziQxdeco&si=LvSUixmnSg4JM1zK
18. https://youtu.be/F21ziQxdeco?si=jyAwbU8dVwkYzRH4

19- https://el-yomnews.com/المكافحة-الحيوية-بالفطريات-حلول-مبتك/

20- https://el-yomnews.com/archives/1296404

21- https://www.facebook.com/share/p/1AC25ue5XG/

22- https://youtu.be/7VItE5kON2g?si=9rMWMUSGk8Csgjfn

23- https://www.youtube.com/watch?v=G9nXA7Q9C60

24- https://www.youtube.com/watch?v=7VItE5kON2g

25- https://el-yomnews.com/archives/1347977

26- https://youtu.be/x9N58DakONg?si=Uh0HbpQsiDRpIGMl

27- https://youtu.be/PfuQy-Zkq3s?si=GtRRxy9XtTW-YgF3

28- https://youtu.be/Vo1NqVmPxMU?si=vJxt4YyhzqJL9hR4

29- https://youtu.be/pqo5F-cbGSc?si=J_B9Dv05bfX5P-7O

TV and Radio

1- Filming in the program (Meeting the Scientists) - Al-Manara Channel, entitled Biological Control of Insect Pests, on 7/2/2013.
2- Filming in the program (Friends of Al-Manara) - Al-Manara Channel, February 2014.
3- A recording of an interview with the program (From the Secrets of the Universe and Life) Radio Sawt Al-Arab, Cairo on 4/26/2015.
4- Filming in the program (Idea for a bobbin) - Nile Cultural Channel under the title "Using entomopathogenic fungi and plant extracts in combating insect pests that infect sugar crops in Egypt" on 11/1/2015.
5- Filming on (This Morning) program - Nile News Channel under the title: Submitting the specifications for the reclamation of one and a half million feddans as a first stage of the reclamation of 4 million feddans on 11/15/2015. https://www.youtube.com/attribution_link?a=NsH2xGv9mY8&u=%2Fwatch %3Fv%3Dsb46rBG8-Mc%26feature%3Dshare
6- Registration in the program (Our Life Between Our Hands) for North Upper Egypt Radio and Radio, presented by Dr. Essam Fathy, 3 episodes, on 11-29-2021, and it will be broadcast on Saturday and Wednesday at 3:30 pm in January 2022
7- Registration in the program (Our Life Between Our Hands) for North Upper Egypt Radio and Radio, presented by Dr. Essam Fathy, 3 episodes, on 11-29-

2021, and it will be broadcast on Saturday and Wednesday at 3:30 pm in January 2022

8- Registration in the program (Our Life Between Our Hands) for North Upper Egypt Radio and Radio, presented by Dr. Essam Fathy, 3 episodes, on 11-29-2021, and it will be broadcast on Saturday and Wednesday at 3:30 pm in January 2022

9- Registration in "Our Green Land" program for North Upper Egypt Radio and Radio, presented by Dr. Essam Fathy, 3 episodes, on 11-2021-2021, to be broadcast in January 2022

10- Registration in "Our Green Land" program for North Upper Egypt Radio and Radio, presented by Dr. Essam Fathy, 3 episodes, on 11-2021-2021, to be broadcast in January 2022

11- Registration in "Our Green Land" program for North Upper Egypt Radio and Radio, presented by Dr. Essam Fathy, 3 episodes, on 11-2021-2021, to be broadcast in January 2022

12- Filming in the program (Egypt is Beautiful) - the second Egyptian Nile Channel, under the title: Climate Changes on November 6, 2022. https://youtu.be/dOujUAZmTdw

13- Registration in "Our Green Land" program for North Upper Egypt Radio and Radio, presented by Dr. Essam Fathy, 3 episodes, on 11-2021-2021, to be broadcast in January 2022

14- Filming in the program (Egypt and Climate Summit) - Delta Satellite Channel under the title: Climate Cop 27 on 13/11/2022 https://youtu.be/Ub8djW0tggo

15- Filming in (Egypt) - Sohag, Coral Times Hotel, under the title: Climate Change, Cop 27, dated 11/15/2022, the journalist Shorouk Abdullah https://m.facebook.com/story.php?story_fbid=492907069288338&id=100076011247165

16- Filming in (Feed and Benefit) - Coral Hotel in Sohag under the title: Climate Cop 27 on 11/15/2022, journalist Ahmed Tarboush https://fb.watch/gPGrDuzZFX/

17- An intervention in the plant clinic program on insect-pathogenic fungi in Egypt's agricultural farms on Thursday 29/2/2024.
https://www.youtube.com/watch?si=4UfyQaeHVLPiVR5J&v=nXV4NT-no7o&feature=youtu.be

18- Filming in the plant clinic program on the Misr Agricultural Channel about insect-pathogenic fungi on Sunday 3-3-2024
https://www.youtube.com/watch?v=sRH44W6TbcQ

19- Filming in the plant clinic program at the Misr Agricultural Channel, early prediction of the red palm weevil and modern methods of combating it, on Thursday 25/4/2024.
https://youtu.be/uq9Qj3AEBHk?si=ymaX06huG2AXJESu

20- Filming in the Green Planet program on the Egyptian Channel 2 (Biological Control) on Sunday 7/6/2024
https://www.facebook.com/watch/?v=981454710052376

21- A press interview in Al-Youm newspaper with journalist Mr. Imad Naseer on 7-30-2024. https://el-yom.com/news/archives/1092767
https://el-yomnews.com/

22- https://el-yomnews.com/?p=1156515

23- https://el-yomnews.com/%D8%A7%D9%84%D8%AF%D9%83%D8%AA%D9%88%D8%B1-

%D8%B9%D8%A8%D8%AF%D8%A7%D9%84%D8%B1%D8%AD%D9%8A%D9%85-%D9%85%D8%AD%D9%85%D8%AF-%D8%B9%D9%84%D9%8A-%D9%8A%D8%AA%D8%AD%D8%AF%D8%AB-%D9%84-%D9%84%D9%84/

24- https://el-yomnews.com/?p=1179515

25- https://www.facebook.com/share/v/4MM85wKRPxsYLqTr/

26- Insects in the Service of Justice: A Lecture Explaining the Role of Forensic Entomology in Investigating Crimes
https://el-yomnews.com/archives/1243894

27- Human health is at risk. An investigation reveals the secrets of pesticide-contaminated agriculture.
https://el-yomnews.com/archives/1296404

28- https://www.facebook.com/share/p/1AC25ue5XG/
The Green Planet Program (Beneficial Insects and Their Role in the Environment) on 5/23/2025 on the Egyptian Channel 2

29- https://youtu.be/7VItE5kON2g?si=9rMWMUSGk8Csgjfn

30- https://www.youtube.com/watch?v=G9nXA7Q9C60

31- https://www.youtube.com/watch?v=7VItE5kON2g

32- https://el-yomnews.com/archives/1347977

33- https://youtu.be/x9N58DakONg?si=Uh0HbpQsiDRpIGMl

34- https://youtu.be/PfuQy-Zkq3s?si=GtRRxy9XtTW-YgF3

35- https://youtu.be/Vo1NqVmPxMU?si=vJxt4YyhzqJL9hR4

36- https://youtu.be/pqo5F-cbGSc?si=J_B9Dv05bfX5P-7O

Dr. Abeer Saleh Alabdallah

PhD in Agricultural Engineering

Department of Plant Protection

Affiliated with the Directorate of Institutes at

Al-Furat University, Syria

Dr. Abeer Saleh Alabdallah, PhD in Agricultural Engineering, Department of Plant Protection, Affiliated with the Directorate of Institutes at Al-Furat University, Syria

Published books: 3

Research papers: 4

Conferences presented: 2

Scientific seminars: 1

Graduation projects supervised

14 years of university teaching experience at the Faculty of Agricultural Engineering, the Faculty of Science in Al-Hasakah, and the Agricultural Technical Institute in Al-Hasakah

Former Director of the Veterinary Technical Institute in Al-Hasakah

Kolibri 360 GmbH
Ferdinand-Jühlke-Straße 7
99095 Erfurt, Deutschland
produktsicherheit@kolibri360.de

Druck:
CPI Druckdienstleistungen GmbH
im Auftrag der
Kolibri 360 GmbH
Ferdinand-Jühlke-Str. 7
99095 Erfurt